Springer
Berlin
Heidelberg
New York
Barcelona
Budapest
Hong Kong
London
Milan
Paris
Santa Clara
Singapore
Tokyo

H. Hoffmeister   M. Szklo   M. Thamm (Eds.)

Epidemiological Practices in Research on Small Effects

H. Hoffmeister   M. Szklo   M. Thamm   (Eds.)

Epidemiological Practices in Research on Small Effects

With 16 Figures and 18 Tables

Contributions based on a conference
held in Berlin/Potsdam from October 10-13, 1995

The conference was organised by the
Robert Koch Institute, Berlin and financially
supported by the Stifterverband für die
Deutsche Wissenschaft, Essen and the VERUM
Foundation – Foundation for Behaviour and
Environment, Munich

This publication was supported by a grant
of the European Commission
(SOC 95 203092 05F03)

 Springer

Professor Dr. Hans Hoffmeister
Robert Koch-Institut
General-Pape-Straße 62-66
12101 Berlin
Germany

Professor Dr. Moyses Szklo
Johns Hopkins University
School of Hygiene and Public Health
Department of Epidemiology
615 N. Wolfe Street
Baltimore, MD 21205
USA

Michael Thamm
Robert Koch-Institut
General-Pape-Straße 62-66
12101 Berlin
Germany

ISBN-13:978-3-642-80465-6

Library of Congress Cataloging-in-Publication Data

Epidemiological practices in research on small effects / edited by
 Hans Hoffmeister, Moyses Szklo and Michael Thamm.
      p.   cm.
   "Contributions based on a conference held in Berlin/Potsdam from
 October 10-13, 1995."
   "The conferenec was organised by the Robert Koch Institute, Berlin
 and financially supported by the Stifterverband für die Deutsche
 Wissenschaft, Essen and the VERUM Foundation - Foundation for
 Behaviour and Environment."
   "This publication was supported by a grant of the European
 Commission."
   Includes bibliographical references (p.).
   ISBN-13:978-3-642-80465-6   e-ISBN-13:978-3-642-80463-2
   DOI: 10.1007/978-3-642-80463-2

   1. Epidemiology--Research--Methodology--Congresses.
 I. Hoffmeister, Hans. II. Szklo, Moyses. III. Thamm, Michael,
 1961-   . IV. Robert Koch-Institut.
 RA652.4.E74 1998
 614.4'072--dc21

Production: PRO EDIT GmbH, D-69126 Heidelberg
Cover Design: design & production GmbH, D-69121 Heidelberg

SPIN: 10630483   27/3136-5 4 3 2 1 0 - Printed on acid-free paper

Contents

Contributors

Ahlbom, Anders
Professor
Institute of Environmental Medicine
BOX 210
S-17177 Stockholm
Sweden

Aoki, Kunio
President Emeritus
Aichi Cancer Center
1-1, Kanokoden,Chikusa-ku
Nagoya 464
Japan

Armenian, Haroutune
Professor
Johns Hopkins University
School of Hygiene and Public
Health
Department of Epidemiology
615 N. Wolfe Street
Baltimore, MD 21205
United States of America

Begg, Colin
Chairman
Department of Epidemiology and
Biostatistics
Memorial Sloan-Kettering
Cancer Center
1275 York Avenue
New York, NY 10021
United States of America

Dickersin, Kay
Assistant Professor
Department of Epidemiology and
Preventive Medicine University of
Maryland
School of Medicine
506 West Fayette Street
Baltimore, MD 21201
United States of America

Doll, Sir Richard
Professor
University of Oxford
CTSU Clinical Trials Service Unit
Nuffield Dept. of Clinical Medicine
UK- Oxford OX2 6HE
United Kingdom

Feinleib, Manning
Research Professor
Institute for Health Care Research &
Policy
Georgetown University Medical
Center
2233 Wisconsin Ave, NW
Washington DC 20007
United States of America

Goodman, Steven
Assistant Professor
Johns Hopkins University School of
Medicine
Oncology Center
550 N. Broadway, Suite 415
Baltimore, MD 21205
United States of America

Hoffmeister, Hans
Professor
Robert Koch-Institute
General-Pape-Str. 62-66
D-12101 Berlin
Germany

Holland, Walter W.
Professor
The London School of Economics
and Political Science
Houghton Street
UK- London WC2A 2AE
United Kingdom

Kohlmeier, Lenore
Professor
University of North Carolina at
Chapel Hill, School of Public Health
2105 E. Mc-Gavran-Greenberg Hall
Chapel Hill, NC 27599-7400
United States of America

Kromhout, Daan
Professor
National Institute of Public Health
and Environment
Antonie van Leeuwenhoeklaan 9
NL- 3720 BA Bilthoven
The Netherlands

Matanoski, Genevieve M.
Professor
Johns Hopkins University
Department of Epidemiology
624 N. Broadway, 280
Baltimore, MD 21205
United States of America

Prentice, Ross
Senior Vice President and
Director of Public Health Sciences
Fred Hutchinson Cancer
Research Center
1124 Columbia Street
Seattle, WA 98014
United States of America

Shapiro, Samuel
Director
Slone Epidemiology Unit
Boston University, School of Medi-
cine
1371 Beacon Street
Brookline, MA 02146
United States of America

Szklo, Moyses
Professor
Johns Hopkins University
School of Hygiene and Public
Health
Department of Epidemiology
615 N. Wolfe Street
Baltimore, MD 21205
United States of America

Überla, Karl
Professor
Ludwigs-Maximilians-Universität
IBE
Marchioninistr. 15
D-81377 München
Germany

Westlund, Knut
Professor
Hamang terrasse 87
N-1300 Sandvika
Norway

Wynder, Ernst L.
President
American Health Foundation
320 East 43rd Street
New York, NY 10017
United States of America

Small effects as a main problem in epidemiology

Hans Hoffmeister, Berlin / Germany

The current development of modern epidemiology has been strongly influenced by the changing patterns of diseases in the western world, namely away from infectious diseases and towards chronic diseases caused by civilization. The famous epidemiological studies, having shown the overwhelming effects of smoking on lung cancer or alcohol consumption in respect to cirrhosis of the liver, as well as having illustrated the concept of cardiovascular risk factors, created the following myth of the sixties and seventies:

Assuming a proper study design and good epidemiological practice, it was hypothesised that there would be a possibility to detect the main causes and prerequisites for every chronic disease.

In case of strong effects leading to relative risks of 4, 5 or more calculated in the classical way (meaning not as a continium in exposed as compared to unexposed populations), the high expectation placed on epidemiological instruments has been proven. It is seldom the case that one or more biases or problems with uncontrolled confounding in a well-conducted and controlled study should produce a correlation despite the fact there is none - or even just the opposite. Influences on health parameters of a certain magnitude will be discovered in any case despite an overestimation or an underestimation of a given risk.

Nonetheless, the belief in epidemiology is changing and a lot of criticism has arisen. This is the case mainly ever since the young science of epidemiology was applied to detect weak associations. A major problem to be solved in analytical and interventional epidemiology of this kind is the minimisation of specific biases and other methodological difficulties inherent in epidemiological work. Is it possible to overcome the problem of bias or is the science of epidemiology now confronted with these methodological limits?

Studies on small effects often produce contradictory results which confuse the public. L. Mayes et al. (1988) reported of more than 50 cause-effect relationships with conflicting evidence, whereby the findings of at least one study contradicted the results of another. In the meantime, numerous additional examples have been reported in the literature. In some cases, the different findings could be explained by taking a closer look into the methods used in these studies as well as in the search for biases, which - in most cases - are responsible for conflicting results. An example might be the story of coffee consumption and blood cholesterol level which illustrates a typical problem with small effects.

E. Bjelke (1974) reported for the first time a strong association between coffee consumption and serum cholesterol in a Norwegian population. Later, in two other Norwegian studies, the same strong effect was observed (Jacobsen and Thelle 1987; Stensvald et al. 1989). Other European researchers, including our group, confirmed these results (Salvaggior et al. 1991; Pietinen et al. 1988; Mensink et al. 1993). However, in similar US-studies, the results were less convincing. In some cases, an effect was observed but not in others (Nicols et al. 1976; Rosmarin et al. 1990).

Based on the importance of US results, there was a lot of scepticism as to whether or not cholesterol levels could really be elevated by coffee consumption. As one can assume, the US studies were biased by something similar to a "negative exposure suspicion bias". However, epidemiologists enjoying coffee in the old and in the new world, will at once taste the "dilution bias" which is responsible for the differences between the European and the US studies. In fact, actually two different beverages were being examined in this discussion: being that US-coffee does not have the same cholesterol increasing effect as a particular Scandinavian coffee.

Meanwhile, the causal agent in coffee, which is primarily responsible for the increase in the serum cholesterol level in a dose response relationship, has been detected. The causal agent is a coffee lipid called caffeol (Zock et al. 1990). It might be added, that even a small enhancement of relative risk for cholesterol by coffee would have a large impact on health since coffee drinking is very common in most western populations. In Germany, coffee consumption raised from 20 l per capita a year in the sixties to 100 l today, causing an important attributable risk.

Furthermore, with regard to the above-mentioned problems involved in proving small effects, A. Feinstein (1988) claimed substantial improvements in epidemiological studies dealing with the menace of daily life. He argued that the epidemiologic methods used may have omitted fundamental scientific standards to specify hypotheses and target groups: get high quality data, analyse attributable action and avoid detection bias. However, I believe the fundamental criticism of Feinstein mainly has nothing to do with omitting scientific standards but with the inherent difficulties entailed when small effects are to be analysed.

Weiss (1990) argued in a contradicting article, that similar methods were used in the studies praised by Feinstein and by those he criticised (both hospital-based and population-based case-control studies and cohort studies on alcohol consumption and breast cancer, as well as coffee drinking and pancreatic cancer). Savitz and co-workers (1990) also showed that Feinstein's accusation against epidemiological methods, standards and studies in general is far away from the reality in this field which is rapidly developing and creating solutions for many of the difficulties arising from weak associations. Nevertheless, modest associations with relative risks close to 1 need to be carefully interpreted. Due to methodological difficulties, these results do not necessarily represent a causal relationship or even a correlation.

In a recent scientific report in *Science* entitled "Sizing up the cancer risks" US epidemiologists were cited as saying that they would not take seriously a single study reporting a new potential cause of cancer unless the exposure to that agent increased a person's risk by at least a factor of 3. Even then, there should be scepticism if the study was not a very large one, was extremely well done or supported in its results by a reasonable biological explanation.

I believe that this can be agreed upon and defines what small effects mean and how they should be ascertained. In the above-mentioned report, 25 cancer risks seen in epidemiological studies and picked up in the popular press over the last years were listed, showing relative risks of not more than 3 to 4. Some of the topics in this list belong to highly-feared issues in the German public, resulting already in increasing political activities and even legislative consequences. Several good examples of these are: long-lasting occupational exposure to dioxin as a risk for all cancers (RR 1.5), pesticide residues in blood as a risk for breast cancer (RR 4; contradicted later in a larger study), exposure to electromagnetic fields and breast cancer (RR 1.4) - not forgetting the never-ending controversy over nuclear power plants and elevated risk of leukaemia in the surroundings of such plants. These topics were recently generated again in one study in Germany but not in others and can be included here as well.

In a recent report in *Science*, the headline stated: "Epidemiology faces its limits; the search for subtle links between diet, lifestyle, or environmental factors and disease is an unending source of fear - but often yields little certainty". This cannot be the last word on the subject.

I strongly believe that analytical epidemiology will have its main tasks in the field of weak associations. Despite the problems of handling small effects, it has to be stated that the epidemiological approach is the only way to confirm or falsify the many concerns about real or assumed risks under living conditions and today's environment. The answer to these and many other questions that epidemiology can give to the existing difficulties is that methodological progress must be achieved, creative solutions found and more sophisticated studies performed.

Biases, confounding factors and other methodological problems do contribute substantially to this uncertainty. They result in a wide range of errors in epidemiological studies dealing with small effects. In the following contributions, these obstacles will be better characterised and ideas will hopefully be developed to minimise them.

4

References

1. Mayes LC, Horwitz RJ, Feinstein AR (1988) A collection of 56 topics with contradictory results. Int J Epidemiology 17: 680-685
2. Bjelke E (1974) Colon cancer and blood cholesterol. Lancet 1: 1116-1117
3. Jacobsen BK, Thelle DS (1987) The Tromsø heart study: is coffee drinking an indicator of a lifestyle with high risk for ischemic heart disease? Acta Med Scand 222: 215-221
4. Stensvald J, Tvordal A, Foss OP (1989) The effect of coffee on blood lipids and blood pressure. Results from a Norwegian cross-sectional study, men and women 40-42 years. J Clin Epidemiol 42: 877-884
5. Salvaggior A, Periti M, Miano L, Onaglia G, Narzorati D (1991) Coffee and cholesterol, an Italian study. Am J Epidemiol 134: 149-156
6. Pietinen P, Geboers J, Kesteloot H (1988) Coffee consumption and serum cholesterol. An epidemiolgical study in Belgium. Int J Epidemiol 17: 98-104
7. Mensink GB, Kohlmeier L, Rehm J, Hoffmeister H (1993) The relationship between coffee consumption under consideration of smoking history. Eur J Epidemiol 9: 140-150
8. Nicols AB, Ravenscroft C, Lamphierer DE, Ostrander (1976) Independence of serum lipid levels and dietary habits. The Tromsek study. J Am Med Assoc 236: 1948-1953
9. Rosmarin PC, Applegate WB, Somes GW (1990) Coffee consumption and serum lipids: a randomised, crossover clinical trial. Am J Med 88: 349-356
10. Zock PL, Katan MB, Mertens MP, van Dusseldorp M, Harranvan JL (1990) Effect of a lipid risk fraction from boiled coffee on serum cholesterol. Lancet 335: 1235-1237
11. Feinstein AR (1988) Scientific standards in epidemiologic studies of the menace of daily life. Science 242: 1257-1263
12. Weiss NS (1990) Scientific standards in epidemiolgic studies. Epidemiology 1: 85-86
13. Savitz DA, Greenland S, Stolley PD, Kelsey JL (1990) Scientific standards of criticism: a reaction to "Scientific standards in epidemiologic studies of the menace of daily life" by A.R. Feinstein. Epidemiology 1: 78-83
14. Anonymous (1995) Sizing up the cancer risks. Science 269: 165
15. Taubes G (1995) Epidemiolgy faces its limits. Science 269: 164-169

Obstacles in evaluating small effects

Moyses Szklo, Baltimore / USA

Although the main focus of this conference is on validity and not precision, data on precision underscore the challenges of detecting small effects. For example, to be able to demonstrate a relative risk of 5 in the context of a case-control study in which the prevalence of exposure is about five per cent, only seventy cases and seventy controls are needed. On the other hand, if the relative risk one is interested in is 1.1, about 35,000 cases and 35,000 controls are needed for the same exposure prevalence of five per cent. Sample size requirements can, therefore, pose a major initiation to the epidemiologic assessment of a small increase in the relative risk.

Why should epidemiologists be interested in "small effects"? The answer is more easily understood when considering the population attributable risk. For example, if the relative risk associated with a given risk factor is 1.2 - which most of us would define as "small effect" - when one-half of the population is exposed, almost ten percent of the disease can be explained and possibly prevented in the target population. Thus, a small "relative risk" effect does not necessarily translate into a small "attributable risk" effect when the exposure is common. It follows that when referring to a "small" effect the epidemiologist needs to specify how "effect" is measured.

An example of the importance of not discounting a small relative risk is exposure to high blood pressure in relation to cardiovascular outcomes. The relative risk for disease such as strokes increases in a graded fashion as the blood pressure levels increase. Thus, compared to low levels of blood pressure, the relative risk is progressively higher as the blood pressure increases. However, when the population attributable risk is examined, it is quite obvious that most cases of cardiovascular diseases originate in persons with mildly, but not those with severely elevated blood pressure. This is because the prevalence of moderate hypertension is much higher than that of severe hypertension. Thus, it is essential to distinguish the relative risk in persons at "high risk" (i.e. those with severe hypertension) from the population attributable risk, which is a function not only of the relative risk but also of the prevalence of the exposure (i.e. moderate hypertension).

Another example of the dichotomy between the relative risk and the attributable risk is renal insufficiency. If high levels of creatinine are used to define renal insufficiency, the relative risk of high creatinine levels increase with increasing levels of both diastolic and systolic blood pressures. In some studies, the relative risk for the upper blood pressure quartile compared to the lowest quartiles is greater than 3.0. However, as shown for cardiovascular outcomes, the largest number of renal insufficiency events originates among persons with relatively small elevations of blood pressure, e.g. diastolic levels of 90-104 mm Hg, usually defined as "borderline" hypertension.

The primacy of the population attributable risk - first described by M. Levin - when the main focus is prevention, explains why seasoned epidemiologists postulate a population-wide approach rather than a high risk approach for prevention. For the specific example of blood pressure or renal outcomes, the former approach would translate into shifting the whole blood pressure curve to the left, so as to prevent the maximum number of cases in the population as a whole.

Another issue relevant to the discussion on small effects is biological plausibility. For example, why are investigators so convinced that lipoprotein (a), a molecule very similar to that of low density lipoprotein that aggregates in families with a history of heart disease, strengthen the notion that the association is causal. Biological plausibility as well as some of the other criteria discussed by B. Hill, such as dose-response, should be carefully taken into consideration when evaluating small effects.

It is also important to take into account what A. Corres calls in his presentation "drowning of susceptibles", that is, the presence of an effect modifier that produces a clear-cut increase of risk in only a small proportion of the population, thereby diluting the average effect for the population as a whole. Thus, for example, the overall association between salt and hypertension would be difficult to detect if only relatively few individuals in a given population were genetically susceptible to salt-induced hypertension.

An additional obstacle to assessing small increases in risk has been addressed by E. Wynder in a recent commentary published in the *American Journal of Epidemiology*. This deals with the problem of exposed controls, underscoring the dependency of etiologic epidemiology on the non-experimental approach and, thus, on the presence of a sufficiently variable exposure within a study population. For example, the unexpected finding that fat intake was unrelated to coronary heart disease in the Framingham study could be explained by the fact that, particularly a few decades ago, the intake was homogeneously high in the U.S. population.

Finally, issues related to bias and confounding are crucial when assessing small effects. For instance, even a moderate misclassification of either exposure or outcome can lead to failure to detect a small relative risk increase. Improvement of the validity of measure of putative risk factors or outcomes is obviously a function of scientific developments in general, thus underscoring the need for close collaboration between epidemiologists and basic scientists.

Towards good epidemiological practices

Ernest L. Wynder, New York / USA

If I had read the much discussed article by G. Taubes (1995) in *Science* entitled "Epidemiology faces its limits" as a young person, I would probably never have entered this field. Yet, I firmly believe that, in the realm of all sciences that are part of the many disciplines of medicine, epidemiology is the key. Having studied demographics, lifestyles, and disease patterns for more than four decades, I firmly believe that most of the chronic diseases of our time are not inevitable consequences of ageing, but rather that they relate to metabolic overload. As M. Skzlo stated in his presentation, we need to define what is "optimal" in terms of physiologic and metabolic norms and what is optimal in lifestyles. A systolic blood pressure of 130 and a serum cholesterol level of 200 mg/dl is not optimal. Clearly, smoking of tobacco or marijuana is also not optimal. If most people had a blood pressure of 110/70, and a cholesterol level of 140 mg/d, and if most adults abstained from smoking, coronary artery disease would be rare indeed. As epidemiologists, we need to recognise that we do not only perform descriptive work but that, on the basis of translational research data coming from environmental chemistry, biochemistry, molecular biochemistry and from various clinical observations and data, we need to be equally involved in the application of preventive strategies and in monitoring the effects of intervention. Those among us who discover risk factors should also make certain that - once identified - such factors are reduced or eliminated.

As a young medical student, when I first wanted to examine whether smoking might cause lung cancer, this subject was certainly not in the forefront of attention in the medical community, as R. Doll will well remember. I guess I was lucky, in the sense that I entered a field that is related to a major risk factor. I had, instead, at that time pursued the impact of hair colouring on cancer in women. I might as well have buried myself in a laboratory, or studied some outlying speciality of medicine. Both R. Doll and I know that the relationship of cigarette smoking and lung cancer was not regarded as a viable concept in the fifties (Doll & Hill 1950; Wynder & Graham 1950). I would like to point out that in my first study, without meaning to insult those among you who are statisticians, that at least 80 percent of the cases were interviewed personally by me. I emphasise this because among young epidemiologists today, interviewing by principal investigators no longer seem to be standard practice. Mistakes may be made due to the fact that certain designated individuals conduct the interview, others input computer print-out. Some of these print-outs certainly look very impressive but they do not always convey all that needs to be known about a case. In those early days in St. Louis,

Missouri, I prepared the flow sheets in my own handwriting - which gives you an idea about my support facilities at the time. However, this practice also allowed me to be very intensively involved with the data and more importantly, to have control over their interpretation. As we had no department of statistics at Washington University at the time, the data were published without statistical treatment. As I said, the strength of these data spoke for themselves and I was lucky that I happened to study a risk factor of such significant magnitude. M. Szklo sometimes quotes me as having said "if a correlation is really significant, you do not need a statistician; and if you do need a statistician, you are probably in trouble".

The observations E. Graham and I published in 1950, were not widely acknowledged in America, certainly not by the tobacco industry, and most certainly not even by the major statisticians of that time, such as Fisher (1957) in England and Berkson (1955) in the United States. It was not until 1962 that the Royal College of Physicians issued a warning, and not until 1964 that the Surgeon General of the U.S. Public Health Service aroused public attention to the health hazards of tobacco smoking (Berkson 1955; US Public Health Service 1958).

Odds ratios established in the early studies on smoking and lung cancer sometimes reached as high as 40:1. Today, epidemiologists are troubled by weak associations or small effects. I can agree with M. Szklo that a weak association of 1.3 - if true, and if affecting a large population - can be of major public health consequence. The problem areas affecting weak associations arise mainly from the way we select our cases and controls and deal with bias, confounders and subgroup analysis (Royal College of Physicians 1992; US Public Health Service 1964; Feinleib 1987).

For case-control selection, let me give two examples of problems affected by confounders. One relates to the study of hair dye which revealed a confounder that was not suspected. Colleagues at New York University published a paper showing that hair dye was a significant cause of cancer of the breast (Harris & Wynder 1988). We took a look at the controls to see who was using hair dye and soon found that there was a high correlation between use of hair dye and being a Jewish woman. Adjustment of the data for religious background of cases and controls ruled out the use of hair dye as a contributing factor to the risk of breast cancer (Wynder 1990). A few years later, when the research group at New York University adjusted their data for religious background among cases and controls, they came to the same conclusion reached by us (Shore et al. 1979).

Several studies showed a correlation between use of mouthwash and oral cancer, particularly for users of Listerine®. I do not know if Listerine® is sold all over the world, but I am told that it tastes so bad that it must be effective. When we closely examined Listerine® users, we found that they also smoked more and drank more than non-users (Wynder & Goodman 1983). The mouthwash helped them to cover up the tell-tale signs of both smoking and drinking. When we adjusted smoking and alcohol consumption, the formerly incriminating variable "mouthwash" disappeared.

A good epidemiologist also needs to have a sense for biological plausibility of any perceived relationship of risk factors to disease. Principally, following Robert Koch's postulates, the full range of criteria of judgement of causality for chronic disease factors were initially introduced by Hill (1957). I have stated them repeatedly (Wynder 1983) and they were restated to clarify the relationship between smoking and cancer in the US Surgeon General's Report on smoking and health (1964). The point I like to re-emphasise is that in studies linking smoking to lung cancer, we found odds ratios as high as 40:1 and yet, the US Surgeon General demanded that the criteria of judgement of causality be applied towards evaluation of these data. Today, investigators who come up with odds ratios between 1.1 and 1.3 generally do not invoke the criteria of judgement of causality. I think one should not accept the validity of a study that reports a weak association without going through that exercise. There are many studies in the literature on the relationship of alcohol use to breast cancer. More of them imply than deny it, but they certainly do not pass the test of the criteria of judgement of causality, especially with regard to ecological distribution (US Public Health Service 1964; Hill 1957; Wynder 1954). Causality in the relationship of an environmental agent with induction of cancer must satisfy a number of requirements. In 1956, I have stated these as follows (Wynder et al. 1983).

1. The relative risk must rise in proportion to the degree of exposure.
2. Rates among specific population groups must be consistent with the distribution of the agent.
3. Withdrawal, reduction, or modification of the agent in a population group must be followed by a decrease in incidence of a given cancer after a suitable latent period.
4. The agent should be shown to be carcinogenic to some animal species.

In 1953, we induced cancer of the skin by painting solutions of tobacco "tar" on the backs of mice (Harris et al. 1988). In 1957, we elicited cancer of the rabbit's earlobe by topically applying tobacco "tar" solutions (Wynder & Harris 1989). Why did we not blow cigarette smoke into a cage with laboratory animals to see if we could induce cancer? Such relatively futile assays were actually attempted by some investigators (Wynder et al. 1957), and yet, from studies of the nasal turbinates of small animals it was apparent that due to being close to the ground for millions of years, they have developed very intricate nasal passages that were efficient filters. Thus, upper respiratory airways retained particles of the smoke that would likely have caused lung tumours if this material had reached the susceptible tissues of the lungs (Wynder et al. 1957). Homo sapiens have been walking erect for millions of years and his nasal turbinates are not quire as intricate; yet, this highly evolved creature makes a deliberate effort to inhale smoke deep into the lungs through the oral passage and upper airways, bypassing whatever defence nature intended against inhalation of particles. Interestingly, to this day we have never seen an animal with the same ambition.

A few years ago, J. Berger and I examined various confounders that were identified among a large number of smokers. The cigarette smoking habit is, of course, highly related to educational status, especially among men, but also among women: It is correlated with alcohol use and coffee drinking. It is less well known that cigarette smoking is also highly correlated with intake of dietary fat, particularly in the fat in meat, and that it is negatively correlated with intake of fruit. Thus, when we see interpretations in the literature leading to the statement that "eating fruit protects against cancer of the lung", we need to cautiously check whether the investigators have well adjusted the data for tobacco consumption. We want to know how early a subject began to smoke and which kind of cigarette and how many of them he or she smoked and how deeply these smokers inhaled. Regarding the latter, we have only recently learned to use computerised techniques to assess individual smoking behaviour and found from these studies that depth of inhalation and total uptake of smoke is always governed by nicotine satiation, so that smokers of low-yield cigarettes are not really protected in a way that one would expect from reading labels that indicate lower smoke yields (Wynder & Hoffmann 1996).

Body weight is often a strong confounder. Among women, weight, in turn, can be heavily confounded by educational status. The more educated women become, the thinner they get. I guess this has to do with their desire to wear designer clothes. The correlation of body weight and smoking is interesting in that it is U-shaped for men and almost that way for women (Berger & Wynder 1994). Persons who are light smokers are thinner. A heavy smoker weighs more, perhaps because he also drinks more alcohol, eats more, and is generally less concerned about the health consequences of his behaviour. Smoking correlates to age. Heavy smokers tend to die prematurely (Berger & Wynder 1994).

Even though the existence of confounders of epidemiologic data has been recognised, confounders have not always been successfully identified. One must think of them and then examine their validity and the degree to which they may have an impact on the interpretation of data. Thus, one must have an appropriate questionnaire. In carrying out a cohort study, or a study such as the one involving diesel engine exhaust exposure in railroad workers, unless good data on tobacco smoking are obtained, one cannot reach a conclusion about diesel engine exhaust as a risk factor (Kabat et al. 1994).

The wish bias is even more difficult to deal with than the confounder. Given the correct hunch about confounders and having gathered the right kind of data, confounders can be adjusted for the wish bias. However, this is a very subjective factor (Royal College of Physicians 1992; Zang & Wynder). For example, about 150 km outside New York City is a popular summer resort called Southampton. It takes about two hours and ten minutes to get there by car. If one lives out there, one gets used to the commute and, if asked how long it takes to get there, one might say "oh, just about an hour and a half". However, if one asks a person who does not have a house there why he or she shies away from renting or buying one, the answer is likely "I do not have a house there because it takes four hours to get there." The wish is also inherently different for cases and controls (Zang & Wynder). One of the wish biases which pose a problem in our current studies relates to answers we got from women, and particularly from women with breast cancer about their fat intake. Over the past four years while our study was in progress, the more the topic of diet and breast cancer appeared in the daily papers or was mentioned in newscasts and discussed on talk shows in the media, the more we found women underreporting their fat intake in a 24 hour diet recall. This poses a considerable problem because, all of a sudden, the women we interviewed tell us that they hardly eat any fat. When we calculated fat intake from the data we have gathered, the figure was at or above 25% of total calories. Another wish bias relates to height and weight. We asked members of our staff to state their height and weight. One week later we actually measured and weighed them. Men consistently overreported their height, while women underreported their weight. We have to acknowledge these wish biases as part of human nature and treat them accordingly in our evaluations.

What about a cohort study? If I am a smoker with lung cancer who works for an asbestos company, or if I have had any kind of occupational exposures, I will certainly not admit that it is my smoking that is responsible for the disease. I rather wish for the culpability of my employer. A lung cancer patient is far more likely to acknowledge exposure to asbestos than to confess his smoking history. In fact, upon being given a diagnosis of lung cancer, one of my patients said "well, I am not surprised because some thirty years ago I worked in a hardware store where a sack of asbestos fell on the ground and I had to clean it up." Surely any patient in the control group is free of such incriminating memories. Then, there definitely is a wish bias for research scientists and we need to acknowledge this too. Researchers would rather report a positive than a negative finding. If they only come up with negative findings, their chances for grant support are slim because funding is more likely reserved for an acute problem that may be resolved by the proposed research. In this case, I agree with G. Taubes (1995) that we scientists have a tendency to overreport positive findings, while the wish bias that is associated with underreporting is part of human nature. When researchers operate under a wish bias, the result is poor science. Unfortunately, we see this not only among epidemiologists but also among our colleagues who conduct bioassays in laboratory animals. Lastly, there is also a wish bias towards positive findings among journal editors and reviewers. One might call this "publication bias".

To provide a baseline for nutrition-related studies, W. Mertz (1991) at the Nutrition Research Center of the United States Department of Agriculture has determined what people had actually eaten and then compared these data to the information given in questionnaires on dietary intake. This showed underreporting of caloric intake by more than 20%. Applying certain urinary markers of food intake such as nitrogen excretion in conjunction with carefully measured food intake, S. Bingham at Cambridge (1982) found that among those reporting their fat intake to be below 60 g per day, everybody underreported. Why is that important? In the study of breast cancer among nurses with W. Willett (1987), the lowest quintile reported eating was 58 g fat or less. Because these women were nurses, we must be prepared to concede that they have probably underreported, because they are certainly more aware of the health risks associated with a high fat intake than women outside the nursing profession.

We need also to concern ourselves closely with the "exposed control group" as a problem in weak associations (Bingham 1982). In respect to odds ratios for smokers, the usual reference group consists of those who have never smoked. However, if we select the reference group to be "those smoking 1-19 cigarettes per day", the odds ratios become smaller. This problem exists also in the evaluation of nutrition-related cancers, especially in homogeneous populations in which the range of exposures is narrow. Most studies in nutritional cancer epidemiology in the Western world suffer this limitation as the range of fat consumption among cases and controls in most studies is too narrow to yield significant differences in relative risks even though the average fat consumption by cases is truly higher than that of controls. Only comparative studies of hugh population samples would reveal this. On the other hand, when fat intake ranges widely, as is the case in comparing cancer epidemiology of Asian and Western populations, significant differences can be more readily discerned. We have also discussed these important issues through correspondence in the *Journal of the National Cancer Institute* (1995) in the context of a well-designed case-control study by A. Whittemore et al. (1995) who examined prostate cancer rates in relation to diet, physical activity and body size among various populations.

I spent many years examining vital statistics from the Japanese people. It is striking that there are enormous differences in the rates of virtually every major type of cancer when we compare the Japanese data with those in the United States. They all seem to relate to lifestyle. Some of the most glaring differences are in the incidence and mortality of cancers of the uterine cervix, breast and ovary. If breast cancer rates are further dichotomised into premenopausal and postmenopausal occurrence, there is an eight-fold difference in breast cancer mortality. Cancer of the bladder is six times less common in Japan than in the United States even though Japanese men smoke more than their American counterparts. Is this a genetic factor? Hardly, because when the Japanese moved to Hawaii, their rate of bladder cancer as well as that of most other cancers goes up (Wynder 1997). On the other hand, the Japanese have by far the highest incidence of and mortality from stomach cancer. As we have stated many years ago, the decline of stomach cancer in the United States is an unplanned triumph and we can only speculate how it happened (Howson et al. 1986). Of course, we prefer to see a decline of cancer rates without knowing exactly why it happens than to witness increases of cancers that are related to a known cause. The epidemiology of stomach cancer is a good example of the idea that cancer is not an inevitable consequence of being alive but rather it relates to environmental factors which we can control, and they are often dietary in nature.

In response to the question whether epidemiology faces its limits we should examine populations with marked differences in risk and try to explain why this is so. Epidemiology is well served if it can be supported by biomarker studies in metabolic epidemiology and molecular biology. We need to recognise and make the best use of the descriptive and translational aspects of this discipline to get to its ultimate goal, the application of new knowledge and the assessment of the effectiveness of our intervention.

We are using all of these aspects in our approach to research on cancer of the breast and cancer of the prostate, trying to explain how dietary fats affect the induction and development of these cancers, and translating our research findings into management of patients to prevent further progression of the disease. D. Rose (1995) at the American Health Foundation has shown that it is not just fat per se, but specifically linoleic acid which has a strong effect on tumour progression in the athymic mouse model.

In terms of applied epidemiology, I want to re-emphasise that epidemiologists must get involved in helping society reduce those factors that have been definitely linked to occupational, or environmental, or nutrition-related cancer risk. Scientists at the American Health Foundation can take a problem from the first epidemiologic clue to chemical and biochemical analysis of the risk factors, and delineating of mechanisms of action in vivo and in vitro with state-of-the-art methodology, including molecular biology techniques. We can also translate the research findings into health education and health promotion and disease prevention programs for the better health of people everywhere. Because we feel strongly about the need for epidemiologists to monitor the effectiveness of the preventive strategies we apply, the American Health Foundation has - as a large scale national trial - initiated the Women's Intervention Nutrition Study (WINS) (1992). The protocol for this study provides for 2,500 American women with stage 1 or stage II breast cancer who are treated with either radiotherapy or chemotherapy, to be randomised into an intervention group on a low fat diet. (i.e. only 15% of caloric intake from fat), and a group without intervention, remaining on the usual diet. To date, about 8,000 women are enrolled in this study in 32 participating cancer centres in the United States. We anticipate that the patients on the low-fat diet will have a better survival rate and fewer recurrences of cancer. Soon we will begin a very similar study among men who, after prostatectomy, develop a very high prostate-specific antigen (PSA) level. These patients will be randomised into the intervention group receiving a low-fat diet, and a second intervention group, receiving a standard diet (30% calories from fat) but, in addition, also a chemo-preventive "cocktail" consisting of selenium, vitamin E, and genistein (a soy pro-

14

tein). In this case, we are fortunate to have the PSA test as a biomarker. The study is designed to show differences in PSA velocity between cases and controls and we are hopeful that it can be controlled through dietary management.

Often, one learns from history; J. Lind (1753) described how scurvy could be prevented when he did not have a clue about vitamin C. Semmelweis' studies (1847) not only provided excellent epidemiologic observations but also experimental evidence for the cause of puerperal fever. Yet, it took a long time for him to convince physicians of the necessity to scrub. We did not need to wait for Louis Pasteur (1879) to identify the bacillus streptococcus to know how to prevent infection with this organism.

At this time, our greatest concern is to find the most effective means to prevent the epidemics of chronic man-made diseases that prematurely kill so many people throughout the world. Long before we understood the infinite details of mechanisms leading to cancer and other diseases, long before we know all about the intricate molecular events that may allow us to alter genes towards resistance to carcinogenic insult, we can take steps to impact cancer risk, cancer incidence and cancer mortality. The aim of epidemiology and of medicine in all its applications is truly to help people die young as late in life as possible. We can help to fulfil this aim by unambiguously identifying strong and weak causative associations and by reducing or eliminating them.

References

1. Taubes G (1995) Epidemiology faces ist limits. Science 269: 164-169
2. Doll R, Hill AA (1950) Smoking and carcinoma of the lung. A preliminary report. Med J 2: 739-748
3. Wynder EL, Graham EA (1950) Tobacco smoking as a possible etiological factor in bronchiogenic carcinoma. J Am Med Assoc 143: 329-336. Also cited and reprinted as "Landmark Research" (1992) in J Am Med Assoc 253: 2986-2994 and in the J NIH Res 4: 63-72
4. Fisher RA (1957) Alleged dangers of cigarette smoking (letter). Brit Med J 2: 1518
5. Fisher RA (1958) Lung cancer and cigarettes? (letter) Nature 182: 108. Also: Cancer and smoking? (letter) (1958) Nature 182: 596 of PD. Stolley. When Genius Errs. Fisher RA (1991) and the lung cancer controversy. Am J Epid 133: 416-426
6. Berkson J (1955) The statistical study of association between smoking and lung cancer. Proc Staff Meetings Mayo Clinic 30: 319-348
7. Smoking and lung cancer: Some observations on two recent reports (1958) J Am Statist Assoc 53: 28-38
8. Royal College of Physicians. Smoking or health? 1992. London, Pitman Medical Publishers.
9. US Public Health Service (1964) Smoking and Health. A report of the Surgeon General. US Publ Hlth Serv Publ No 1103, Washington, DC. US Govt. Printing Office.
10. Feinleib M (1987) Biases and weak associations. Prev Med 16: 150-164

11. Harris RE, Wynder EL (1988) Breast cancer and alcohol. A study in weak associations. JAMA 259: 2867-2871
12. Wynder EL (1990) Epidemiological issues in weak associations. Int J Epidemiol 19: 5-7
13. Shore RE, Pasternack BS, Thiessen EU, Sadow M, Forbes R, Albert EA (1979) Case-control study of hair dye use and breast cancer. J Natl Cancer Inst 62: 277-283
14. Wynder EL, Goodman MT (1983) Epidemiology of breast cancer and hair dyes. J Natl Cancer Inst 71: 481-488
15. Koenig KL, Pasternack BS, Shore RE (1991) Hair dye use and breast cancer: a case-control study among screening participants. Am J Epidemiol 133: 985-995
16. Wynder EL, Kabat G, Rosenberg S, Levenstein M (1983) Oral cancer and mouthwash use. J Natl Cancer Inst 70: 255-260
17. Hill AB (1957) Smoking and cancer of the lung. Lancet 2: 1289
18. Wynder EL (1954) Tobacco as a cause of lung cancer. The Pennsylvania Med J 57: 1073-1083
19. Harris RE, Spritz N, Wynder EL (1988) Studies on breast cancer and alcohol consumption. Prev Med 17: 676-682
20. Wynder EL, Harris RE (1989) Does alcohol consumption influence the risk of developing breast cancer? Two views. Important advances in oncology. DeVita Jr. VT, Hellman S, Rosenberg SA. Eds. pp. 283-293: Philadelphia, J.B. Lippincott.
21. Wynder EL, Graham EA, Croninger AB (1957) Experimental production of carcinoma with cigarette tar. Cancer Res 13: 855-864
22. Graham EA, Croninger AB, Wynder EL (1957) Experimental production of carcinoma with cigarette tar. IV. Successful experiments with rabbits. Cancer Res 17: 1058-1066
23. Wynder EL, Hoffmann D (1967) Tobacco and tobacco smoke. Studies in experimental carcinogenesis. Chapter 7: 206. New York, Academic Press
24. Berger J, Wynder EL (1994) The correlation of epidemiological variables. J Clin Epidemiol 47: 941-952
25. Djordjevic MV, Fan J, Ferguson S, Hoffmann D (1996) Self-regulation of smoking intensity. 1. Smoke yields of the low-nicotine, low-tar cigarettes. Carcinogenesis 16: 2015-2021
26. Kabat JC, Chang CJ, Wynder EL (1994) The role of tobacco, alcohol use and body mass index in oral and pharyngeal cancer. Int J Epidemiol 23: 1137-1144
27. Zang E, Wynder EL (submitted for publication) Lung cancer and alcohol consumption. A study in confounding
28. Muscat JE, Wynder EL (1995) Diesel engine exhaust and lung cancer: an unproven association. Environm Health Persp 103: 812-818
29. Wynder EL, Higgins IT, Harris RE (1990) The wish bias. J Clin Epidemiol 43: 619-621
30. Mertz W, Tsui JC, Judd JT, Reiser S, Hallfrisch J, Morris ER et al. (1991) What are people really eating? The relation between energy intake derived from estimated diet records and intake determined to maintain body weight. Am J Clin Nutr 54: 291-295. Also: Mertz W (1992) Food intake measurements: Is there a "gold standard"? J Am Dietetic Assoc 92: 1463-1465
31. Bingham S, Wiggins HS, Englyst H, Seppanen R, Helms P, Strand R et al. (1982) Methods and validity of dietary assessments in four Scandinavian populations. Nutrition and Cancer 4: 23-33

16

32. Willett WC, Stampfer MJ, Colditz GA, Rosner BA, Hennekens CH, Speizer FE (1987) Dietary fat and the risk of breast cancer. New Engl J Med 316: 22-28
33. Wynder EL, Stellman S (1992) The overexposed control group. Am J Epidemiol 135: 469-461
34. Whittemore AS, Kolonel LN, Wu AH, John EM, Gallagher RP, Howe GR et al. (1995) Prostate cancer in relation to diet, physical activity and body size in blacks, whites and Asians in the United States and Canada. J Natl Cancer Inst 87: 652-66. See also Wynder EL, Stellman SD, Lumey LH, Winters B, Cohen LA (1995) Correspondence re the above article (1995) J Natl Cancer Inst 87: 1329
35. Wynder EL, Hirayama T (1977) Comparative epidemiology of cancers of the United States and Japan. Prev Med 6: 567-594
36. Howson CP, Hiyama T, Wynder EL (1986) The decline in gastric cancer; an unplanned triumph. Epidemiol Revs 8: 1-27
37. Rose DP, Connolly JM, Liu XH (1995) Effects of linoleic acid and -linoleic acid on the growth and metastasis of a human breast cancer cell line in nude mice and on its growth and invasive capacity in vitro. Nutrition and Cancer 24: 33-45
38. Chelebowski RT, Rose DP, Buzzard IM, Blackburn GL, Insull Jr W, Grosvenor M, Elashoff R, Wynder EL (1992) Adjuvant dietary fat intake reduction in postmenopausal breast cancer patient management. The Women's Intervention Nutrition Study (WINS). Breast Cancer Res. Tretm. 20: 73-84 (Review)
39. Lind JA (1753) Treatise on the scurvy. 486 p. London, England, A. Milard.
40. Semmelweis I (1847) Höchst wichtige Erfahrungen ....epidemischen Puerperalfieber. Zeitschr. Ges Wien 4: 242-244
41. Pasteur L (1879) Septicemie puerperale. Bull Acad Med Paris 2nd Series 8: 256-260

Problems in detecting small effects in case-control and cohort studies

Daan Kromhout, Bilthoven / The Netherlands

As you all know, a publication in *Science* some months ago dealt with the limitations of epidemiology (Taubes 1995). In that publication it was said that epidemiology is left to studying weak associations. The big question is: Is that statement true? I would like to give a few of examples of weak associations. But before that, I would like to question the thesis that we are only studying weak associations. Within a couple of weeks we will have a publication in the *British Medical Journal* showing a risk ratio of 15 (van Asperan et al. 1995). So it does not mean that we are only left with risk ratios in the order of 1 - 2. Sometimes also strong associations can be found today, as was shown in a study we carried out last year in The Netherlands.

July 1994 was the warmest month of July in the Netherlands since 1706. At my institute we were notified that there was an outbreak of otitis externa in the eastern part of the country. We were asked to investigate the cause of that outbreak. We figured out that the people who got otitis externa were swimming in sweet-water lakes. We took watersamples in those lakes. Smears were also taken of the ear and a bacteria, pseudomonas aeruginosa, was identified as the causal agent. The risk ratio between the exposure to the bacteria pseudomonas aeruginosa and the occurrence of otitis externa was 15.

I would like to start with two statements from an article on limitations of epidemiology in *Science* (Taubes 1995). The first one is from M. Angell, one of the editors of the *New England Journal of Medicine*. In her article she said that "As a general rule of thumb we are looking for a relative risk of 3 before accepting a paper for publication." The second statement is from R. Temple of the US Food and Drug Administration. He said "My basic rule is: if the relative risk is not at least 3 or 4, forget it." The question is: Are these statements true? I do not mean true in the sense of whether they are right or wrong, but whether one can make this type of absolute statements. Is it justified? I would like to show you some examples in my presentation that question this type of statements. The reason is that, as always, in epidemiology the situation is very complicated. Several questions can be asked, for instance in relation to exposure assessment. My first question is: How good can we measure the exposure of interest? Is the measurement of exposure a good reflection of the true exposure? For example:

Take for instance the association between beta-carotene intake and lung cancer. There are a lot of case-control studies showing an association between the consumption of certain fruits and vegetables and the occurrence of lung cancer. Some scientists deduced from these results that beta-carotene could be the most important nutrient in those fruits and vegetables in relation to lung cancer. If you take, however, the evidence from cohort studies - as we have done recently - then the evidence for an association between beta-carotene intake and lung cancer is much weaker (Ocké et al.). Recently we published results based on the Seven Countries Study where we actually measured carotenoids in food composites, representing the average food intake of 16 cohorts (Ocké et al. 1995). These food composites were chemically analysed and different carotenoids were related to 25-year mortality from lung cancer. We did not find an association between carotenoid intake and lung cancer mortality. You may remember the results from the ATBC trial in Finland, where they supplemented smokers with beta-carotene and vitamin E (ATBC Study Group 1994). In that trial no protective effect of beta-carotene on lung cancer was found. On the contrary, it was even the other way around. Do we then really know what we should measure? Meaning, when we are estimating the consumption of fruits and vegetables, is it correct to take a food table and to translate foods into nutrients and to say that the exposure of, for instance beta-carotene, is responsible for the protective effect of some fruits and vegetables? In a recent publication in Lancet, it was shown that the absorption of carotenoids is an extremely complicated issue (De Pee et al. 1995). We do not even know what type of carotenoids are easily absorbed and what types are not. Measurement of exposure is an important and complex problem.

The second issue that I would like to address is the problem that in many epidemiological studies only one measure of exposure is taken. Generally, we take a baseline measurement and follow-up on participants for quite a long time in prospective studies. We then calculate the association between one baseline exposure measurement and an outcome variable. Is that the proper way of identifying relationships? Like the group of R. Peto has shown several years ago, for the relationship between blood pressure and different cardiovascular outcomes (Mac Mahon et al. 1990), repeated measures are needed to adjust for regression dilution. This is also complicated if one is for instance interested in diet and cancer relations. If dietary measurement are repeatedly done using the same method, then normally correlations in the order of 0.8 are to be found (Bloemberg 1989). In this case, the real question is: are we not repeating the same error at the same time? Do we then really measure the true exposure? We do not know.

Finally, we have to address the issue whether the observed association is plausible? I would like to give an example by referring to an article published by us a couple of years ago on the association between bird-keeping and lung cancer (Holst 1988). The group of H. Hoffmeister and L. Kohlmeier from the Robert Koch Institute also worked on this issue (Kohlmeier et al. 1992). I started the research because a general practitioner came to my office and said that he had been a GP for about ten years and thirteen of his patients have died from lung cancer. He observed that all of these patients kept birds. In his opinion, bird-keeping is an important determinant of lung cancer. I found the story so strange that I did not believe it in the beginning. I argued with the GP that this association was confounded by smoking, and smoking is the most important risk factor for lung cancer. The GP convinced me finally to do a case-control study. We did the study and found a risk ratio of 5 for the occurrence of lung cancer when we compared patients who kept birds with those who did not. In this case, the issue is: do we have a plausible explanation for that association? It can be argued that small dust and psittacosis infection could play a role. But is that really true? How much do we know about the possible causality of this association?

I would like to give you two examples of associations that have been established. I think that few epidemiologists will question the causality of the associations between blood pressure and stroke and between serum cholesterol and coronary heart disease. What I would like to illustrate is that even in such situations the question of small relative risks crops up. It is therefore not always the case that when we talk about causal associations, we also talk about large relative risks. The first example concerns the association between blood pressure and stroke. I will illustrate with data from the Zutphen Study. Zutphen is a small commercial town in the eastern part of The Netherlands. It is a town with about thirty thousand inhabitants. It has only one hospital and it is a beautiful place for doing epidemiologic research. The study was started in 1960 as a Dutch contribution to the Seven Countries Study. At that time there were about 2,500 men in Zutphen aged 40-59 years. A random sample of about 900 men was taken from the population. That sample has been followed-up since that time.

In the first part of the study, each year the men were examined by a team headed by Van Buchem, the first principal investigator of the Zutphen Study. I took over from him in 1977. The 15th round was carried out in 1977/78 and the 16th round in 1985. We continued thereafter with the 17th round in 1990, the 18th round in 1993. Last Spring, the 35-year follow-up survey was conducted. An enormous amount of information on this group is available.

Today I will only use data up till 1985. Yearly blood pressure measurements were done between 1960 and 1970, and 15-year incidence of stroke from 1970-1985. This made it possible to analyse the influence of repeated measures on the strength of the association between blood pressure and stroke. In the first analysis, blood pressure measurements were taken in 1970 and related to stroke incidence during the next 15 years. We can say this is ordinary epidemiology. Simply take one measure of exposure and relate that to the outcome of, for instance, a period of 15 years. In the second analysis, the blood pressure measurements from 1960 to 1970 were regressed on time and then calculated in 1970 from the regression line. In the third analysis, only the average blood pressure over the 10-year period was taken and then related to stroke incidence over 15 years. This would give an idea on the order of magnitude that the risk ratio is changing by using different operationalisation of exposure measurements.

The results were published some years ago (Keli et al. 1992). These are as follows: 557 men without strokes were selected and compared with 46 men who had strokes. A difference can be seen when we take the 1970 blood pressure measurement of about 7mm of mercury in systolic blood pressure between the cases and the non-cases. This difference is slightly larger, but not much larger, when we take the average blood pressure from 1960 to 1970 and compare no-stroke cases with stroke cases. An interesting phenomena is, that the standard deviation becomes smaller if you go from one to repeated measures. This means that the estimate of the blood pressure measurement is becoming more precise when using repeated measures. The true association is therefore much stronger than the observed association when casual blood pressure measurements were used.

The data was also analysed in another way. This is also reported in (Keli et al. 1992). For example, men with a systolic blood pressure above 155 were compared with those with blood pressures below 130. If the observed blood pressure in 1970 is taken, a risk ratio of 1.84 is observed. Due to the relatively small group, this risk ratio is not statistically significant. When the predicted systolic blood pressure of 1970 in the two groups was compared, the risk ratio was observed to be 2.34. But when the average of 11 measurements was used, the risk ratio increased to 3.11, and is statistically significant. This analysis shows a strong influence of the amount of information that is available on the exposure. It is therefore very important for epidemiological studies not to take only one exposure measure. Repeated measures are important in order to get an insight into the true strength of the association. We also calculated the increase in the strength of the association by taking those repeated measures and came up with about 55%. In the publication by Peto's group (1990), where a different approach was used, regression dilution in the association blood pressure and stroke is also of the same magnitude. This means that the strength of repeated measures in relation to blood pressure and stroke is about 50% larger, taking regression dilution into account as compared to the situation where this is not done.

A second example to be presented here is about the association between serum cholesterol and the occurrence of coronary heart disease. This example is based on the Seven Countries Study, initiated by A. Keys in the late fifties. The seven countries involved were the United States, Finland, the former Yugoslavia, Japan, Greece, Italy and the Netherlands. A total of 16 cohorts were involved. In the former Yugoslavia there were even five cohorts. In the United States and the Netherlands, there was only one cohort. In all the other countries, at least two cohorts existed. An analysis on the association between serum cholesterol and the occurrence of coronary heart disease was recently carried out and published by us (Verschuren et al. 1995). Some results from that publication are as mentioned below.

The study was a prospective cohort study. The total number of cohorts, as mentioned earlier, was sixteen. The total number of men was about 13,000 aged

40-59 years at baseline. They have been followed now for 25 years. The baseline surveys were between 1958 and 1964. The baseline survey was repeated after 5 and 10 years, and the mortality follow-up was continued till 25 years. Thereafter, the study was terminated. The vital status of all the participants and of the 13,000 at risk was checked. Only about 50 were lost. The vital status after 25 years was thoroughly checked and the total number of deaths was about 6,000 during this period.

In this analysis, 16 cohorts were taken. The number of men involved in the different cohorts varied from 500 to 2,500. A cohort of 500 men is relatively small. The 16 cohorts were regrouped into six homogeneous groups. The first group was Northern Europe, including Finland and the Netherlands with a total number of about 2,500. The cohort of US railroad workers had also about 2,500 men and the Southern Europe inland group about 3,000. The latter group consisted mainly of the cohorts from the former Yugoslavia and inland Italy. In Southern Europe, there were four real Mediterranean cohorts. They were all along the coast and grouped together. The two Serbian cohorts Zrenjanin and Velika Krisna were excluded from inland Europe because they had very large increases in serum cholesterol levels over the first ten years of follow-up. There was a 30% increase between 1960 and 1970 in average serum cholesterol level, and therefore they have been put together as a separate group. Lastly, a follow-up on about 1,000 persons was done in Japan. Around 1960, large differences in average cholesterol levels within Europe were observed. In Northern Europe, it was 6.5 mmol/l in contrast to an average level of only 4.2 mmol/l in Serbia. This picture has completely changed over the last 25 years with simular levels observed all over Europe. During the 25 years of follow-up there was also a large difference in mortality from coronary heart disease. It was only about 5% in the Mediterranean cohorts and more than 20% in Northern Europe.

We were interested in the relative risk and absolute risks in the different cultures. A. Keys showed in his monograph (1980) that a strong association exists between serum cholesterol and coronary heart disease in the United States and Northern Europe, and probably no association in Southern Europe. However, mortality from coronary heart disease in Southern Europe was much lower than in Northern Europe. After ten years of follow-up, the number of CHD cases was too small for a proper analysis of that association to be made. Therefore, this was done for only six different groups (Verschuren et al. 1995). In all groups with the exception of Japan, there is an association between serum cholesterol and mortality from coronary heart disease. The relative risks in those cohorts were similar but the absolute risks differed greatly. Taking a serum cholesterol level, for instance, of about 5.5 mmol/l it can be seen that in the Japanese and in the Mediterranean cohorts only about 3% of the men died from coronary heart disease during the 25 years of follow-up. In Northern Europe, however, this is about 15%. Thus at the same serum cholesterol level there can be a very large difference in the absolute risk for coronary heart disease, even when the relative risk is the same.

22

The question of regression dilution was addressed. We used the baseline and the 5-year follow-up data and compared the extreme quartiles to calculate the regression dilution factor. For four of the six groups, a regression dilution factor of about 1.4 was found, but the regression dilution factor was lower in Serbia. This was due to the large increase in the cholesterol level in Serbia over 10 years of follow-up. A factor of about 1.4 has also been published by M. Law about a year ago (Law et al. 1994). In three publications on serum cholesterol and coronary heart disease, the regression dilution issue was also addressed and a dilution factor of about 1.4 was found. It can be concluded that, if population serum cholesterol level is stable, a dilution factor of about 1.4 is observed. In the situation of an increasing population serum cholesterol level, the dilution factor is confounded by the change in cholesterol level. It is therefore not possible just to calculate general dilution factors, but the dynamics of the biological risk factors in populations have also to be taken into account.

Even in the last example showing the association between serum cholesterol and coronary heart disease, the risk ratio, if comparing extreme quartiles of exposure, is only about 2. This is in a situation where only one measurement exists. If a correction for regression dilution can be done, then a risk ratio of about 3 will be obtained. Even then, we will be in the area of low relative risks. Of course, the risk ratio can be increased by including more extreme parts of the cholesterol distribution. However, it would need very large groups at risk to get stronger associations.

I would like to conclude that absolute statements about the relative risk level in relation to causal inferences cannot be made. One has always to make up one's mind and not to jump simply to conclusions based only on a risk ratio. To judge the causality of associations with a low relative risk, information is needed from other observational studies in any case. No definite conclusions can be made if the results from only one study are available. Associations can only be judged causal when results from observational studies are backed up by the results from experimental and clinical studies.

References

1. Taubes G (1995) Epidemiology faces its limits. Science 269: 164-169
2. van Asperen IA, Rover CM, Colle C, Schijven JF, Bambang Oetomo S, Schellekens JFP, van Leeuwen WJ, Collé C, Havelaar AH, Kromhout D, Sprenger MJW (1995) Risk of otitis externa after swimming in recreational fresh water lakes containing Pseudomonas aeruginosa. Brit Med J 311: 1407-1410
3. Ocké M, Bueno de Mesquita B, Feskens E, Van Staveren W, Kromhout D (accepted) Repeated measurements of vegetables, fruits, and antioxidant (pro-) vitamins in relation to lung cancer (The Zutphen Study). Am J Epidemiol
4. Ocké M, Kromhout D, Menotti A, Aravanis C, Blackburn H, Buzina R, Fidanza F, Jansen A, Nedeljkovic S, Nissinen A, Pekkarinen M, Toshima H (1995) Average

intake of antioxidant (pro) vitamins and subsequent cancer mortality in the 16 cohorts of the Seven Countries Study. Int J Cancer 61: 480-484

5. The Alpha-tocopherol Beta Carotene Cancer Prevention Study Group (1994) The effect of vitamin E and beta carotene on the incidence of lung cancer and other cancers in male smokers. N Engl J Med 330: 1029-1035

6. De Pee S, West CE, Mukalil-Karyadi D, Hautvast JGAJ (1995) Lack of improvement in vitamin E status with increased consumption of dark green leavy vegetables. Lancet 346: 75-81

7. Mac Mahon S, Peto R, Cutler J et al. (1990) Blood pressure, stroke and coronary heart disease. Part 1. Prolonged differences in blood pressure: prospective observational studies corrected for regression delution bias. Lancet 335: 765-674

8. Bloemberg BPM, Kromhout D, Obermann-de Boer GL, Van Kampen-Donker M. (1989) The reproducibility of dietary intake data assessed with the cross-check dietary history method. Am J Epidemiol 130: 1047-1056

9. Holst PA, Kromhout D, Brand R (1988) Pet birds as an independent risk for lung cancer. Brit Med J 297: 1319-1321

10. Kohlmeier L, Arminger G, Bartholomeyczik S, Bellach B, Rehm J, Thamm M. (1992) Pet birds as an independent risk factor for long cancer: Case-control study. Brit Med. J 305: 986-989

11. Keli S, Bloemberg B, Kromhout D (1992) Predictive value of repeated systolic blood pressure measurements for stroke risk. The Zutphen Study. Stroke 23: 347-351

12. Verschuren WMM, Jacobs DR, Bloemberg BPM, Kromhout D, Menotti A, Aravanis C, Blackburn H, Buzina R, Dontas AS, Fidanza F, Karvonen MJ, Nedeljkovic S, Nissinen A, Toshima H (1995) Serum cholesterol and long-term coronary heart disease mortality in different cultures. Twenty-five-year follow-up of the Seven Countries Study. JAMA 274: 131-136

13. Keys A (1980) A multivariate analysis of death and coronary heart disease. Cambridge, England, Harvard University Press 1-381

14. Law MR, Wald NJ, Wu T, Hackshaw A, Baily A (1994) Systematic underestimation of the association between serum cholesterol concentration and ischaemic heart disease in observational studies: data from the BUPA Study. BMJ 308: 363-366

Proposals and recommendations concerning small effects in case-control and cohort studies

Karl Überla, Munich / Germany

The main conclusion of D. Kromhout in detecting small effects is that a causal relationship can exist when relative risks are smaller than 3. Therefore, an absolute estimate or absolute statement on risk ratio levels in relation to causal inference cannot be made. I cannot share his point. It does not seem logical to me. The existence of false negative and false positive results in low risk associations does not prove that there is no formal relation between risk ratio levels and causal associations. At higher risk levels, the number of false negatives and false positives might be smaller than at lower risk levels. The existence of a few examples is logically not sufficient for such a strict conclusion. I think we are empirically quite sure that the chance of a casual association is higher when you have a risk ratio of, say, 10 or 15 than a risk ratio of 1.5. Therefore, I disagree with D. Kromhout's strict formulation. On the contrary, empirical observations point clearly to the fact that - as indicated in the *Science* article - relative risks smaller than 3 are usually controversial for a long time. A strong relationship between causal inference and high relative risks cannot be excluded by some false negative or false positive examples in low risk associations.

I want to add some examples from my experience which relate to the topic. They are not widely known. The first approach relates to the way in which risks are presented. I call this the risk pictogram. It was published in 1990 in a supplement to the *International Journal of Epidemiology* with the title "Boundaries of Perception and Knowledge of Risk Assessment in Epidemiology" (Überla 1990). I would like to explain this approach briefly.

In epidemiology we know of four measures of risk which are interrelated. The incidence in the control group or population I_0; the incidence in the exposed group I_I; the relative risk, which is the ratio o the I_I to I_0; an the attributable risk, which is the difference of I_I minus I_0. We have to consider them simultaneously. The relative risk and the attributable risk have different meanings. Our risk measurements are connected by definition. They are not independent.

26

Figure 1
Risc-Pictogramm

Relative Risc RR and incidence I_0 in their relation
to the Attributable Risk A

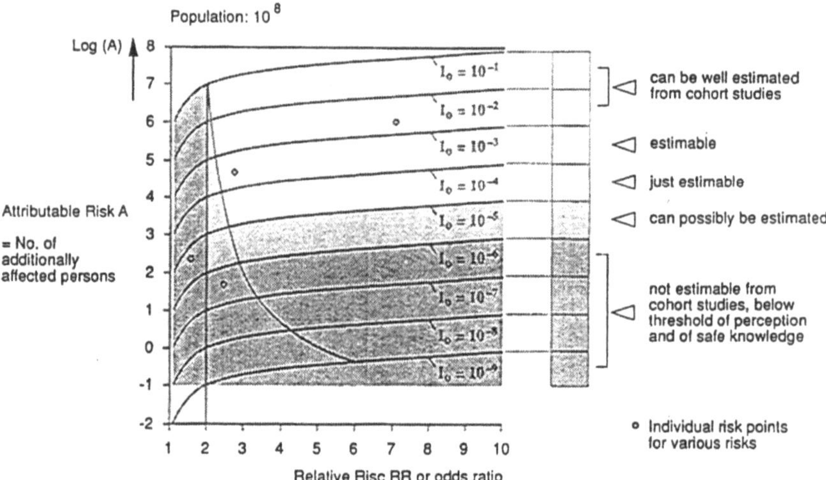

I have proposed a unified way of presentation in a single graph containing all
the possible information on epidemiologic risk measurements (Figure 1). To the
right the relative risk or the odds ratio is scaled. The vertical axis shows the attributable risk in a log scale, that is the number of additionally affected persons. In
order to get an attributable risk, one has to refer to a certain population with a
definite number of persons. I selected size simply by shifting the scale up and
down. The curves show the mathematical relation between the incidence in the
population I_0, the relative risk and the attributable risk. For instance, when one has
a relative risk of 3 and a basic incidence I_0 in the population which is not exposed
of 10^{-5}, then the attributable risk on a log scale is around 3, which means that in a
population of 100 million there are 1.000 people additionally affected by the risk.
When the relative risk and the incidence in the population I_0 is known, it is possible to look up immediately to the left what the attributable risk is. There are two
thresholds of perception, which are indicated by the grey area. The first is the
incidence in the control group. A risk of 1:1.000 can easily be estimated in cohort
studies. However, when there is a risk of 1:1.000.000 this is not possible. There is
an empirical boundary below which a low risk cannot be established quite correctly. Nobody can follow a cohort of 1 million for 10 years without a loss of information. Therefore, we are not in a safe position to establish such a low incidence.

We are below the threshold of perception and safe knowledge from epidemiologic evidence. Our formulas and theories are correct to any decimal points but the empirical data limit the evidence more than we like.

There is a second threshold, which is somewhere at relative risk levels of 2 or 3. As pointed out in the *Science* article (Taubes 1995), when the relative risk is smaller than 2 or 3, this is outside the area where we can infer safe knowledge from epidemiology. The third curve in the middle of the graph is somehow arbitrary. It describes the plausible fact that with decreasing incidence I^0 the border-line between safe knowledge and guesswork should shift to the right. For every single study it is possible to place a point in Figure 1 as indicated. Such a point can be in the safe area, or outside safe knowledge. Confidence limits can be attached to these points. Such a pictogram can be modified in various ways. It is a unified way of presenting risks, including all information: the relative risk, the attributable risk and the incidence in the control population. It can be seen also immediately whether a study lays in a safe area; whether or not one can safely infer from epidemiologic evidence. Various risks can be easily compared.

My second proposal is the construction of causal scores. The criteria for judging causality of associations first published by B. Hill have been widely used. They have been criticised by the notion that causal inference is not a matter of science. I think that the concept of causality is very important for our problems and criteria for causality could be revived. I have proposed formalising such criteria using scores (Überla 1990). This is done in medicine in other areas with some success. The easiest way to do this is to take the ten wellknown questions which are used in epidemiology to assess causality of association (Figure 2).

Figure 2
Construction of Causal Scores

Questions included in causal score	Weight in total score[a]
Undoubtful consistency an replicability	1
Exposure measure reliable and validated (r >.7)	1
Outcome measure reliable and validated (r > .7)	1
Main bias factors sufficiently excluded	1
Main confounding factors sufficiently excluded	1
Statistical significance p< .05 twosided	1
Strength of association: RR greater than 2 -3	1
Dose-response-relationship: p < .05 twosided	1
Intervention effect shown	1
Biologic plausibility likely	1
Maximum total score	10

[a]other weights are possible

One can answer YES or NO to any of these questions. If the answer is NO, it counts 0, and if the answer is YES, it counts 1. Then the total causal score varies between 0 and 10. All the questions have equal weights in Figure 2. Different weights could of course be used. One could apply all the instruments for constructing scores to devise different causal scores. When such a score is used, it could be evaluated in various ways, for instance in the way indicated in Figure 3.

Figure 3
Evaluation of the Total Causal Score

Total causal score	5: No indication of causal connection
Total causal score	6: Some weak hint for causal connection
Total causal score	7: Causal connection possible
Total causal score	8: Causal connection likely
Total causal score	9/10: Causal connection existent

When the value is smaller than 5, there is no indication of a causal connection. When the score meets 9 - 10, all those criteria are fulfilled. A causal connection seems to exist. Often the value will be somewhere between and should leave the judgement open. When such scores are applied in low-risk associations, it ends with much more conservative results compared to looking at meta-analysis or at a single study. One could apply various available statistical instruments for the construction of scores. This could help to separate predictive and causal epidemiologic knowledge from less predictive associations.

The third approach uses the incorporation of information to the risk measure from the population, regarding the heterogeneity of strata. It is wellknown that the Mantel-Haenzel chi as well as the Mantel extension chi assume homogeneity across strata, which is rarely present. In two papers with Ahlborn and Tutz (1988, 1990) we have proposed to incorporate information about a cofactor in the population. Doing this changes the risk measures. We proposed a class of chi-square distributed global measures to test dose-response relationships. In contrast to the Mantel-Haenzel extension, chi homogeneity of risk is not assumed. I will not go into further details here. I think that integrating information from the population to the estimation of risk could improve the estimation of small risks when heterogeneity of strata is present.

The funding of epidemiology and its impact on societies is based on the credibility of our results. With inconsistent, irreproducible and instantly changing results, epidemiology will loose its credibility as a science. There is no commonly accepted scientific way to separate safe knowledge from guesswork when the risks are small. This will last for a long time to come. However, we are able to establish boundaries of perception and knowledge in epidemiology. There are such boundaries in other sciences as in physics, in chemistry or in toxicology. We also have to establish such thresholds in order to cope with the irritating situation described by the recent article in *Science* (Taubes 1995).

Finally, I have some recommendations to make:

1. Let us forget relative risks or odds ratios smaller than 2 or 3. They are generally below the threshold of perception.
2. Let us present risks in a standardised way indicating whether they are serious or not.
3. It is necessary to compare risks. This can be done in a standardised way.

Let us concentrate on the larger avoidable risks. There are not so many of them.

References

1. Überla K (1990) Boundaries of perception and knowledge for risk assessment in epidemiology. Int J Epidemiol 19 No. 3 Suppl. 1: 81-83
2. Taubes G (1995) Epidemiology faces its limits. Science 269: 164-169
3. Ahlborn W, Tuz HJ, Überla K (1988) Estimating relative risks from heterogeneous strata. Inf Med 27: 118-124
4. Ahlborn W, Tuz HJ, Überla K (1990) Risk analysis in cohort studies with heterogeneous strata. A global Ch^2-test for dose-response-relationship, generalising the Mantel-Haenzel-procedure. Meth Inf Med 29: 113-121

Comments on problems with small effects in case-control and cohort studies

Knut Westlund, Sandvika / Norway

In his contribution D. Kromhout discussed a phenomenon to which we may return repeatedly in the following, namely regression dilution or, as we used to call it in the old days, regression towards the mean. I met this as an analytical problem in 1967, when J. Cornfield and W. Haenszel gave a course in cancer epidemiology in Oslo. Among their exercises was one on the relationship between serum cholesterol and coronary heart disease. The simplest way to adjust, if you have a linear relationship and can approximate the intraindividual variation, is to turn the slope. With our Norwegian data from around 1960 we ended up with roughly a 40% increase in the relative risks. However, it is not easy to measure such an increase directly. It is more difficult today than in the past. D. Kromhout spoke of taking the average of 11 consecutive measurements. In Norway in the seventies and eighties we have been conducting very large cardiovascular surveys, in which 22,000 men and 22,000 women were examined twice at an interval of 3-5 years, with a 3-9 year mortality follow-up after the second screening. The improvement in the slope by using the mean of the two predictor measurements, rather than a single one, turned out to be much less than prescribed in theory. With such a long interval between the measurements they no longer represent random samples from a constant level - quite apart from the attention given by clinicians to the top measurements, which may well distort any risk-curve more seriously than was the case 30 years ago. Also, there is a difference between the size of the relative risk one needs in establishing a new association and the risks that is met in subgroups when elaborating well-known causal relationships. Of particular importance is the relative risk in old age. At age above 70 there is certainly a coronary heart disease relative risk above 1 associated with the upper quartile of both systolic blood pressure and serum cholesterol. But those relative risks are small, probably less than 2. The absolute rate differences, however, are larger than at younger ages. This is where the pharmaceutical companies come in. The findings in controlled trials in middle-aged males are extrapolated to old men and women. Should otherwise healthy old persons with serum cholesterol about what is now the age and sex specific mean in Norway be put on cholesterol lowering medication? What harm this might do, I do not know. But it certainly costs money. When I try to speak against it, I am told that my job as an epidemiologist is to describe the risk relationships. Let others more competent assess the relative importance of relative and attributable risks.

Small effects and the selection of study participants in case-control and cohort studies

Haroutune K. Armenian, Baltimore / USA

This presentation is based on the principle that epidemiology is an information science. Data generated in epidemiologic investigations need to be used for decision making. Within such a process of decision making, it is much easier to make inferences when dealing with large effects. Such large associations may be identified by any professional with some rational judgement. One has to remember the number of times where the initial association between disease and aetiology has been established by a good clinician. One of the more recent instances of establishing such an association is the example of the eosinophilia myalgia syndrome as a result of ingestion of tryptophane. The initial association was established first by a general internist in New Mexico in spite of a long list of differential diagnosis for such cases of eosinophilia. Difficulties with making inferences arise when we are dealing with small effects. That is when there is dependence on epidemiological expertise. This is when the "trained eye" or mind of an epidemiologist is needed.

Within the process of selection of study participants, this presentation will try to answer the following two questions:

1. What are some factors that lead to small effects?
2. What can be done to prevent the influence of these factors?
   These issues will be discussed using the following outline:

1   Determinants of small effects in the selection of study participants in

1.1 General
1.2 Cohort studies
1.3 Case-control studies

2   Measures to prevent small effects in

2.1 General
2.2 Cohort studies
2.3 Case-control studies

3   Conclusions

1 Determinants of small effects in the selection of study participants

34

1.1 General

Problem definition. The definition of a problem is the first step of any epidemiologic investigation. Our ability to be specific in our definition of the problem reflects the level of our knowledge about the problem. Problem definition is also influenced by our ability to intervene and affect the course of the problem.

Problems where the definition is too diffuse and non-specific will end up with a smaller effect estimate because of strong potential for misclassification. These are problem definitions which are made out of the congruence of multiple issues. For example, a study on the relationship of high cholesterol levels and cardiovascular diseases in general (including cerebrovascular accidents and aortic aneurysms) diseases may lead to a smaller estimate of effect compared to a study where such a relationship is being studied with a more specific disease entity such as coronary artery disease.

Definition of outcome. The definition of the problem delineates our definition of outcome of interest. The international classification of diseases uses both etiologic and manifestation criteria for defining diseases. The more certain we are about cause(s), the more etiological the disease definition would be.

A disease classified by etiology is a disease where the causal relationship is very specific and strong. Thus, the association between the vibrio and the disease cholera is stronger and more specific than the association of a rhinovirus and common cold. Cholera is a disease classified or defined by an etiological factor while common cold is a manifestational entity where the disease diagnosis is not dependent on the presence of a single etiologic factor.

Diseases classified by etiology have at least one strong association, while manifestational entities like cancer and cardiovascular disease reflect the fact that not a single etiologic determinant of these conditions is identified. For most factors, we are dealing with weak associations.

Over the years, epidemiologists have started their outbreak investigations by formulating an epidemiological definition of a case, which is an effort of defining the case by some known characteristic of all cases of interest. In the original investigation of the outbreak, the epidemiological definition of a case of Legionnaires' Disease was formulated as those patients with pneumonia who were at a particular hotel where most of the cases occurred during the epidemic. This step of developing a definition of the outcome based on some common characteristics of the known case is an effort at providing more specificity and hopefully at increasing the chances of identifying a stronger association. In many outbreak investigations there is a continuous reassessment of the definition of a case, as additional evidence of factors involved in the aetiology and transmission are identified. Such an approach is dictated by the pressures of problem-solving in an acute event situation.

Selection of base population. How well the problem is circumscribed to its universe of impact will influence the specificity and the strength of the relationship that is observed. Identifying the cohort or the group within which the particular effect of the exposure is delimited, provides a base population where effect would be amplified. In most investigations, one needs to start by maximising the possibility of identifying a larger relative risk. If we need to study the relationship of uranium mining to lung cancer, then it is better to do that study in a base population where a large number of uranium miners live. In such a population the opportunity for exposure is higher and the probability of detecting a large association larger.

1.2 Cohort studies

Types of cohort studies. As will be discussed later, cohort studies are not the best method to study small effects, but within cohort studies, the best seems to be the non-concurrent ones.

Misclassification of outcome. Misclassification is the most important common cause of artifactual small effects in epidemiological studies. The non-differential misclassification would tend to direct associations towards the null or a relative risk of one.

Latency bias. This is the situation where a cohort study is designed in such a manner that it does not encompass the full period of latency or incubation of the development of the disease. The observed association during the early phases of the incubation period or latency may be small or non-existent.

For example, a study on leukaemia in Hiroshima and Nagasaki in 1947 may have shown at most a small effect of radiation since the minimal latency of radiation-induced leukaemia is about two years. Similarly, one may miss an effect, or identify a small one, if the association that is being studied is conducted beyond the upper bounds of the distribution of the incubation period.

1.3 Case-control studies

Selection of cases

Problem definition. As stated previously, the problem definition determines definition of the outcome or cases.

Depending on the problem that we are interested to study, our selection of cases would vary and our definition of the case will change. In a study on the outbreak of cholera in Peru, the investigators were interested in the problem of the causes of transmission of the disease outside the household or home. Thus, they decided to define their cases as those of first occurrence in the family in order to study exposures outside the household.

Case definition. The specificity of case definition as derived from a definition of the problem of interest affects our ability to identify associations. Better validation of the cases will minimise misclassification and possibly improve the probability of finding a strong association.

Severity and pathology. Certain pathologic tissue characteristics may be related to particular types of exposure. A study that analysis the data by pathologic subtypes or subgroups of severity may identify more specific associations.

Prevalence incidence bias. If exposure influences the duration of the disease following exposure, then selecting prevalent rather than incident cases in a case-control study may influence the odds ratio in a direction that is dependent on the effect of the exposure on the duration.

Selection of controls

Level of generalisation. Depending on the level of generalisation one needs to make from a case-control study, one may choose a different type of control group. Such a choice may influence the level of comparability achievable between cases and controls.

Comparability-matching. By establishing a high degree of comparability, either through individual or group matching, we are hopefully circumscribing the study to a population base where it is relevant to study this association.

Misclassification - subclinical cases. Controls that are subclinical cases may lead to misclassification and an odds ratio that is closer to one.

Study design - incidence density. A discussion of controls will have to take into consideration the different types of design in case-control studies. For example, a case-control study where the cases are selected using the incidence density sampling method may have a higher probability of misclassification of cases as controls.

2 Measures to prevent small effects

2.1 General

Specificity of problem definition. One may try to be more specific as to the problem under investigation. One approach to develop more specific problem definition is to break the problem into its subcomponents.

Define a subcohort through the outcome definition. Use the outcome definition to circumscribe the most efficient sub-cohort or base population within which your investigation should be conducted.

2.2 Cohort studies

Cohort studies are not the best design for studying small effects. In addition to minimising misclassification in both outcome and exposure measurements, one may consider non-concurrent designs or include a follow-up for the full latency period of the disease under consideration.

2.3 Case-control studies

During case selection:

1. Study subgroup or subtype of cases
2. Make case definition more specific
3. Validate the diagnosis in a subgroup

During control selection:

1. Select the controls from the same base population as the cases
2. Improve comparability of the controls to the cases
3. Use multiple control groups

3 Conclusions

Within the framework of selection of the study population, one needs to consider that small effects may be due to:

1. The nature and reality of the association
2. Misclassification due to a number of artefacts and problems of selection
3. Confounding
4. Interaction that is antagonistic

We need to be careful about our design and analysis to avoid artificial association and misclassification. The effect of potential confounders can be dealt with at the design as well as the analytic phases of our investigation. The possibility that small effects we are observing may be the result of an interaction of an antagonistic type has to be studied through appropriate analysis.

Commentary on small effects and the selection of study participants in case-control and cohort studies

Walter W. Holland, London / UK

I have great difficulties in following H. Armenian because he has dealt with most things that are necessary. So I intend to be rather more anecdotal and try to give some examples. I will try to be a discussant, and not show slides or overheads because it is more appropriate to consider the issues in more general terms.

What K. Überla said was very important, if we are really concerned with what to do as a result of a finding of an effect. We are dealing with intervention, i.e. public health significance, how to change the risk and try to improve our knowledge base to make better decisions. We would all like to design our studies in such a way that none of the investigations lead to doubt of what can be done. Whether the risk be large or small, we want a study that shows clearly whether there is an effect and thus whether and what intervention is needed. That of course is very difficult, if not impossible.

In the previous presentation, there was concern with the definition of the numerator. The problem in epidemiological studies is that one is concerned with denominators, and the difficulty is to know the factors that influence that denominator. In studying small effects, a major confounder is the composition of the population studied. The individuals that we study are always, or virtually always, "survivors". That is to say they have survived a variety of different past events to reach a given age. It is essential in epidemiological studies, whether case-control prevalence studies, or longitudinal (cohort) studies, to have some appreciation of the past events which could have influenced this ability to be included in studies. Let me give you three examples.

The first example is a study in a developing country on the effect of indoor air pollution, where the hypothesis tested by the investigators was whether indoor air pollution was responsible for childhood respiratory illness. Both the cases and the controls were drawn from the same neighbourhood. The cases were children who had been admitted with pneumonia in the local hospital, and the controls were children who were not ill from the same neighbourhood. Now of course, this is a nonsense design because that particular neighbourhood was heavily polluted from both outdoor as well as indoor pollution from fires and industries. So the indoor air pollution was only a very small part of the total pollutant load to which the cases and controls had been exposed. So it was not particularly surprising that the authors were unable to show any relationship, and thus neglected to emphasise that pollution might be an important factor in developing countries in the causation of acute respiratory illness in the first year of life.

The second example is a very old one. It is a study on post office and telephone workers in the United Kingdom, the United States and Japan. We showed differences in the frequency of respiratory and cardiovascular disease of these workers in the three countries. When individuals aged forty to sixty-four were examined, the question was posed as to how representative were these post office and telephone workers in the three areas studied. Fortunately, there were adequate sickness absence, mortality in service and retirement records. Thus it was possible to examine the health status on entry into the occupation. We could also examine the mortality and sickness experience of all workers in these occupations and industries in the period before they were examined by us, either through premature retirement, through mortality while in employment, or through sickness absence. We were able to confirm that the area with the highest prevalence of respiratory disease also had the highest sickness absence rate from respiratory disease in the previous years as well as the highest frequency of mortality in service from respiratory disease. The same was true for cardiovascular disease. Thus the conclusion of the relative frequency of respiratory and cardiovascular disease in the three countries was strengthened through the knowledge of other data rather than purely from the prevalence investigations.

The third example is more recent. I have been concerned with the illness experience and health service usage of doctors in the United Kingdom. I was concerned that doctors were not receiving good medical care. Doctors have the third highest SMR for suicide. We undertook a small prevalence investigation, in which we questioned about 800 doctors in three districts about their illness experience and the services they received in the previous year. We used management consultants of the same age and sex as a control group. We reckoned that management consultants have similar terms of employment as doctors, and are also exposed to stressful situations. They have both trainees and consultants as well. We found, not surprisingly, that the relative frequency of illness was not different between these two groups. We found very different behaviours in terms of their use of health facilities, of going to the doctor, of being admitted to hospital, etc. The doctors underused and did not seek appropriate care compared to the management consultants. However, there were no major difficulties in the interpretation of this study. The doctors were all interested in our investigation. They knew why we were doing it because they had experienced the problems themselves. We had a response rate in the three districts of somewhere around 85 to 90 per cent. We were reasonably confident that we had a representative representation of illness experience and utilisation of health services in the doctors. However, among the management consultants we only had a response rate of about 67 per cent. The personnel departments refused to identify the individuals to whom the questionnaires were sent, so it was not possible for us to follow up with a second or third questionnaire to get a better response rate. The only way we could cope with that and to find out whether the results were reasonably robust, was to get the sickness absence experience of the group who responded compared to the group who did not respond. The personnel records of the 67 per cent who responded were no different from those who failed to respond to our questionnaire.

I am concerned about the reliability of epidemiologic results in terms of problems of non-response. When we concern ourselves with trying to measure the exposure of a particular population to a particular environmental hazard, we have great problems which we tend to underestimate. L. Kohlmeier is emphasising the problem of time. It is actually very difficult in modern society to obtain adequate methods of measurement of exposure to a particular environmental agent, whether it be environmental or other, over a long period of time because of the greater mobility and the greater changes in society that are now occurring. We thus tend to use different strategies and I fear that we tend also not to be particularly imaginative in the strategies that we use. An example is in the studies of air pollution. We tend to study adults, and yet integration of findings are difficult, because of confounding variables such as smoking habits, occupation and length of time within an area. The study of children is easier. They do not smoke; they are exposed to their parents' tobacco, and they usually do not move around much in the first few years of life. But more importantly, when they are in primary school, they usually live within a one-mile radius of the primary school. Therefore, if one measures the level of air pollution at the primary school, a reasonably good assessment of the level of air pollution to which the children are exposed at least in the year of study is available.

Similarly for example, the question of exposure to lead in the children's environment arises and causes many of the investigations to use blood lead. One of the problems of using blood lead are the variability of the measurement - getting an adequate sample (minute-to-minute variations etc.). A far better method is to take the deciduous teeth that normally fall out from the child, and slice it up appropriately and measure the lead in those teeth. This actually gives you some measure of the total lead exposure over the lifetime of the child before the tooth drops out. Obviously, there are problems in terms of different ages when teeth drop out, variations between social class and so on, which need to be taken into account. Nonetheless, it is a more accurate method of lifetime exposure to lead than a single point measurement.

Finally, I want to deal with the problems of bias and how one gets around it in outcome measurements. Too few of us tend to take care in trying to assess the sources of variability that exist before a study is undertaken. We rely upon diagnoses, on records in case histories, on death certificates for case-control, prevalence or even longitudinal studies. We neglect the preliminary studies that are necessary to look at the variability of these matters. An example of variability was a study of mortality within the European Union. Before studies on avoidable mortality in Europe were done, we performed a study on the variability of mortality certification within the European countries. We took a series of standard case histories agreed upon by physicians in each of the member countries. Ten case histories were then mailed to a random sample of 75 general practitioners who had certified deaths in the previous month. We paid them the same amount of money that they would receive for filling in a real death certificate. They were asked to complete the death certificate for the patients whose case histories they had received as they would normally do. We had a response rate of around 85 per cent in each one of the member countries. These death certificates were then coded locally, centrally, and by the international centre. As a result it was possible to get some idea of the variability of certifying practice within the European Union. This gave us some idea of the sort of differences which would be significant if found in mortality rates between the member countries. You will not be surprised to hear that we did find some differences. The differences were of the order of 20 to 40 per cent in the death certification practices for the same condition between the member countries. However, the differences that one finds for individual causes of avoidable mortality between and within the member countries is six- to tenfold. Thus it is unlikely that the differences found for variations in avoidable mortality can be explained only by differences in diagnostic or certifying practice.

I have tried in this brief discussion to highlight many of the matters that H. Armenian pointed out that are important in trying to design studies better to help decision-makers arrive at sensible decisions.

There are two further points I would like to take up. The first is that I think one of the failures of epidemiology is that it does not make use of enough variability of information. If you take into account that histological diagnoses can have a wide variation, that is you can be absolutely certain and can get agreement from three or four interviewers as to the meaning of a particular slide and on the next slide only from two out of four and on the third slide only from one out of four, this means that you would have a degree of relationship which one might consider whether there is a dose-response relationship. We tend not to use observations sufficiently in epidemiology. We believe that diagnoses are "hard" data although in actual fact they are rarely so.

My second point is that we are naive about physics and chemistry. A biochemical result is just as subject to variation and error as a questionnaire. Unless we make certain that we have a proper method of measurement, understand and know about the degree of variation and reliability, we would be foolish to rely on the results from a laboratory.

I am disappointed that we are so narrow in our approaches as we are too defensive about epidemiology. We do not appear to appreciate that when we concern ourselves with small effects which may be important, it is unlikely that epidemiology alone can provide the entire answer. To emphasise M. Feinleib's point, we do not work well enough with other disciplines, whether they be psychologists, biochemists. economists or sociologists. For example, there is public concern over cot deaths or sudden infant deaths in children in the United Kingdom. A television programme claimed that it was due to the release of either stibine or phosphine which are released from the fire retardants in the cot mattresses. It is unlikely that a clear positive or negative answer can be obtained from epidemiology alone. This requires the co-operation of organic and inorganic chemists, microbiologists and mycologists to develop models to determine whether the claim is feasible. We must work together with these other professionals rather than try to do it all on our own.

Confounding: Its role in weak associations

Genevieve Matanoski, Baltimore / USA

Introduction

Weak associations will be increasingly more important in epidemiology as we try to sort out the etiologic pathways of common but multifactorial diseases. The strong risk factors for disease such as smoking or asbestos and lung cancer have been identified through epidemiology and these situations may be less common in the future. This does not mean that some exposures today may not be causing high risks in the population. These factors are yet to be recognised. However, the usual situation is that epidemiology will be challenged to help the clinicians sort out the risk factors for common diseases where multiple etiologies and multiple steps in the pathway to disease exist. Each factor will then demonstrate only a low risk associated with disease. The challenge will be to determine whether these associations are real. Such data may always be subject to the interpretation of the individual investigator. However, it would be helpful to establish some guidelines that will minimise the chances of identifying inappropriate or false associations and clarifying the true associations. Currently, a weak association is often followed by studies which gather similar information on bigger populations or by doing meta-analysis on multiple populations. These procedures could simply root an incorrect observation in the trappings of validity by making it statistically significant. The observation may still be the result of inadequate attention to the possible confounding variables that are associated with the disease. If we, epidemiologists, cannot set some guidelines for investigations of weak associations which are common characteristics of studies involving multifactorial low risk factors, then we will need to leave the studies of disease etiology to the clinicians who examine the pathologic specimens. Epidemiologists are still working with the old infectious disease paradigm which looks for a single cause for a disease and corrects for strong and obvious confounding variables such as age and sex. This thinking worked for lung cancer and smoking. We need a new set of guidelines and a new mind set for the multifactorial, weak associations which will be common in the future.

Definition

Many epidemiologists have allowed the definition of confounding to become very broad. In fact, if anyone wishes to refute the results of a study, the usual criticism offered is that the study did not take into account some "confounding factors" even though the critics are usually unable to describe what those factors

46

are. Rothman's textbook (1986) offers a clear definition of confounding as shown in Table 1.

Table 1
Definitions of Confounding

To be confounding, the extraneous variable must have three characteristics:

1. "A confounding variable must be a risk factor for the disease"
2. "A confounding variable must be associated with the exposure under study in the population from which the cases derive"
3. "A confounding variable must not be an intermediate step in the causal path between the exposure and the disease"

From Rothman - Modern Epidemiology

He has included the two characteristics noted in most texts, namely that the confounding factor must be related to both the disease and exposure of interest, and it adds the third important characteristic that the variable must not be part of the causal pathway to disease. These principles are often overlooked in describing the confounding variables for analysis of studies. The investigators add a long list of variables to models as "potential confounders" or other risk factors and excuse this procedure with the explanation that we do not know, nor have we postulated a mechanism in the etiology of the disease and so, any factor could possibly be an important confounding variable. For example, if smoking is related to cardiovascular disease and coffee drinking is associated with smoking, then studies might suggest that coffee drinking is also associated with cardiovascular disease. However, obviously, no correction for potential confounding from coffee drinking should be made unless this characteristic has an independent association with the disease. Otherwise, we may weaken the association of smoking and the disease. Consider the more difficult situation related to various components of diet and disease outcomes. The dietary components are generally interrelated and we do not know which ones are associated with a specific disease. How does an investigator decide when to control for other components in examining a specific exposure if we know little about the etiology and do not approach the problem with specific hypotheses to be tested?

Another apparent association which is not a confounding variable occurs when the variable is part of the causal pathway. This may be difficult to recognise. Problems are increasing in this regard as epidemiologists start to examine more biomarkers for disease; subdividing the components of exposure as diet, and determining the interrelationships between physiologic changes and disease and between different disease outcomes. Greenland and Robins (1985) (Figure 1) have diagrammed several pathways of interrelationships of two factors and their association with disease.

Figure 1
Path diagrams illustrating causal structures in source populations.

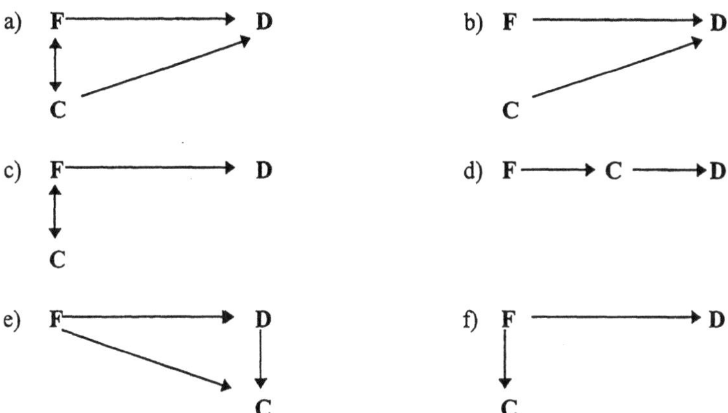

American Journal of Epidemiology 1985;122:496.

In a study, any of these relationships would appear to have an association between the exposures and disease. However, only (a) is a situation in which the factors are truly confounding. Correcting for the second factor in the other situations might dilute an apparent association, especially if the factor was part of the chain of causality. The pathways represent very common situations which occur in investigations of multiple chemicals in industry or relationships between one disease and another. Suppose for example that you wish to study the relationship between rheumatoid arthritis and cancer. How would you determine the role of anti-inflammatory agents in this association if you do not know the pathway to disease and whether most patients with the disease will take the drugs? Epidemiologists must be careful to develop their hypotheses in advance of the analysis and consider carefully the biologic mechanisms in setting these hypotheses and relationships to other factors. Other factors may be discovered as part of the investigation but the authors should clearly identify them as an additional observation.

Confounder influence on weak associations

While epidemiologists recognise the absolute necessity of controlling for confounding factors in studies, they often do not recognise that indiscriminate correction for factors that are not true confounding variables or that are part of a chain in the pathway of disease may reduce the apparent risk associated with the exposure of interest. With increasing accessibility to computers with great speed and memory and to computer programs which provide ease of analysis, there is a tendency to add multiple variables to models in which case the true risk ratios may be diluted. Weak associations are likely to occur when:

1. there is correlation between the variables of interest

48

2. the variables are subsets of a strong risk factor
3. the variables included are part of the causal chain
4. there is unbiased misclassification of exposure variables

Misclassification will be addressed in another paper at this symposium. The other three situations are likely to arise if epidemiologists begin to examine more and more variables in analysis and to examine subsets of suspected risks.

In the past, confounding variables were usually considered to be age, race, gender and poverty as measured by education. These example variables were generally strong confounding factors in the risk of disease. They were also not subjected to major problems of misclassification. But any of these variables is actually a complex mix of multiple potential confounding factors. Age in industrial settings is tied to many important time and dose variables. These include age at first exposure, age attained, duration of exposure, birth cohort, accumulated dose of a chemical, time to tumor and others. If several of these variables are put in the model with age, each may have a low relative risk. However, these subdivisions may indicate important steps in the aetiology of disease. Another example is given in Table 2.

Table 2
Associated with strong factor
Steps:

Poverty (E) risk for disease X

Subdivide poverty e.g. diet (C_1), occupation (C_2), residence (C_3)

What is link?
- Direkt risk from these factors
- Related to exposure factor independently

e.g.

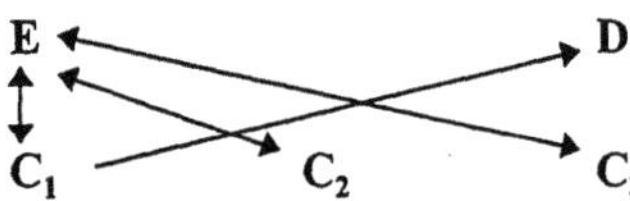

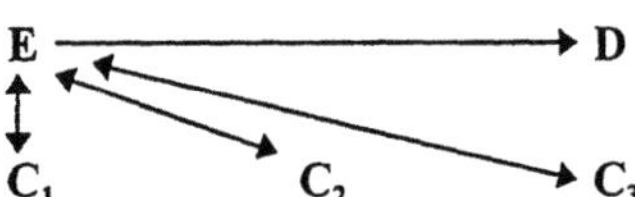

Poverty may not be the true etiologic factor for disease but related risk factors may be the important ones for disease. The example shows a subdivision of poverty into diet, occupation and residence, all factors which may differ by social class, but all these factors may not be true confounding variables in relation to each other. Each may be part of a complex of factors related to disease. Each individual variable may represent a lower risk. Even if these factors were related to disease, it may require several of the factors to produce disease. If any of these situations hold, then using poverty as a risk factor might produce a stronger association than each alone. Yet the subsets may provide more meaningful information on points at which we could intervene. If all factors were related to the disease and were interrelated, then controlling for any, as confounding variables, may dilute the risk still further. Therefore, these subgroups in analysis are important steps for further examination of the etiologic factors. They will not replace age, race, and sex as confounding factors but the direction we are taking is to pursue attempts to move analysis closer to the actual risk factors including biologic markers which are associated with disease. As we do, we must recognise that the factors may not always become stronger risk factors as might have been predicted but results may depend on their relationship to the cause of the disease. We must also be cautious in how we handle these multiple factors in analysis, since interrelationships do not always signify factors that need to be controlled as confounding.

Factors influencing effect of confounders on true relative risk

There are at least three factors which influence the size of the effects of confounders on true relative risk of the exposure under investigation. These factors include:

1. the distribution of the confounder
2. the size of the relative risk of the confounder
3. the degree of misclassification of variables

Ahlbom (1992) has developed a formula which helps to estimate these effects (Table 3).

Table 3
The determination of magnitude of confounding versus main effect exposure

	Confounder	
Exposure	Present	Absent
Present	$RR_{exp} \times RR_{conf} \times I$	$RR_{exp} \times I$
Absent	$RR_{conf} \times I$	I

50

A multiplicative effect between confounder and exposure is assumed since that is the situation in which the RR_{exp} is constant across strata and it is meaningful to talk about one confounding free effect estimate.

Let p be the proportion with the confounder among the exposed and r the same proportion among the unexposed. The magnitude of the confounding may then be taken as:

$$\frac{RR_{obs}}{RR_{exp}} = \frac{p \times RR_{exp} \times RR_{conf} \times I + (1-p) \times RR_{exp} \times I}{r \times RR_{exp} \times RR_{conf} \times I + (1-r) \times RR_{exp} \times I}$$

$$= \frac{p(RR_{conf} - 1) + 1}{r(RR_{conf} - 1) + 1}$$

RR_{obs}	=	Rate ratio with confounding
RR_{exp}	=	Rate ratio for comparison with and without main effect exposure
RR_{conf}	=	Rate ratio for comparison with and without confounding exposure
I	=	Incidence rate with no exposure confounding or main effect
p	=	proportion confounder among exposed main effect
r	=	proportion confounder among unexposed main effect

American Journal of Industrial Medicine 1992; 21:112

This formula indicates that the observed relative risk in the presence of both the main effect and the confounder will differ from the true relative risk of the main effect, depending on the magnitude of the relative risk of the confounder, as well as the proportions of the confounder in both the exposed and unexposed populations. Obviously, the observed value could be increased or decreased depending on whether the confounding variable increases or decreases the risk of disease. This fact is an important consideration since in most situations attention has usually been directed only at confounders which increase the risk and therefore would overestimate the observed value. There are known factors which can reduce risks of some diseases such as drugs and diet. These may be confounders which would result in an apparent weak association which might have been stronger had the confounding association been recognised. On the other side of the coin, if the main effect results in an apparent weak association, but the exposure is associated with a confounding variable with a strong relative risk such as smoking, the entire observed association may be the result of the confounding factor, especially when that factor occurs with high frequency in the exposed population.

This formula helps focus attention on the importance of a differential frequency of occurrence of the confounding variable in the exposed and non-exposed populations. In selecting study groups, epidemiologists should pay careful attention to the confounding variables of age, race and gender, so that the distribution of these factors will be balanced and cannot influence the results. Yet, they will frequently stress the need to control for smoking in occupational studies of two job groups without thinking of whether these groups have different distributions of this confounding factor. Epidemiologists must develop ways to determine or estimate these differential proportions of confounding in exposed and unexposed to find out which factors must be controlled because of their impact as a confounding variable. Adjustments for confounding variables may not always require information on the factor for the total population. For example, in a recent occupational cohort study, we were concerned that smoking might be different in workers compared to the general population used as a comparison. Therefore, we randomly sampled a proportion of the workers and compared their smoking history to that of the population in the National Health Survey (NHIS) (Table 4).

Table 4
Comparison of proportion (%) of ever smoking or chewing tobacco products between the Lifestyle Survey and 1991 National Health Interview Survey
White male

	Living as of 12/31/91		1991 NHS	
	<65	65+	<65	65+
	%	%	%	%
Ever smoke cigarettes	62	77	56	70
Ever smoke pipe	18	32	14	32
Ever smoke cigar	21	29	15	27
Ever chew tobacco	18	18	11	13
Ever use snuff	15	11	9	6

The smoking rates were slightly higher in the workers, so the data were used to adjust the SMR for lung cancer (although the rate of lung cancer was not elevated in the population). In adjusting the SMRs to show the lung cancer risk from working in the industry controlled for smoking, we were able to examine the risks based on different assumed relative risks for smoking and lung cancer (Table 5).

Table 5
SMRs for lung cancer mortality adjusted for proportion of cigarette smoking from the Lifestyle Survey. For different relative risks of lung cancer mortality associated with cigarette smoking. Long-term population, 1970 start, white male

Relative Risk	Adjusted SMR	95% C.I.
2	0.77	0.71, 0.83
5	0.75	0.69, 0.81
10	0.74	0.68, 0.80
15	0.74	0.68, 0.80
20	0.74	0.68, 0.80
Unadjusted SMR	0.80	0.74, 0.86

Higher relative risks result in greater adjustment of the SMR. In this case, however, no further changes resulted after a risk of 10. In this case, differences in smoking rates of six to seven percent are accompanied by a very small change in the SMR, even at high relative risks for the confounder. Epidemiologists should direct more effort at determining the potential influence of various confounding variables on observed associations to see if the differences in exposure to those variables are large enough to influence the apparent weak association which may have been observed.

A very common situation which deserves consideration is that of factors which are highly correlated. We do not often know the risks associated with either of the correlated variables under study. Often, 70 or 80 percent of the exposed population may have the exposure to the possible confounding factor. This situation is common in industry where two chemicals are often closely tied together in a process, or in dietary data where food intakes result in common combinations of nutrients. If we examine the risk of disease from the suspect confounder, it may be high because of the association with the exposure of interest. This situation can present a major problem. If we control for the suspect confounder, the risk associated with the variable of interest may be weakened. In fact, the important exposure may be the combined exposure or some additive effect of each of the exposures. These types of problems should be carefully examined to try to identify the important exposure pathways before being simply satisfied with an adjustment for all variables but the one of interest.

Multifactorial causal chain or confounding

Epidemiologists are focusing more attention on the investigation of factors directly associated with the risk of disease. This could result in including multiple independent risk factors for disease in a model and treating the variables as if they were confounding. Such a situation might result in confusion in regard to the risk from any one of the variables. An example of such a combination of factors is presented in Table 6. The investigation includes three independent risk factors for a disease such as a toxic exposure, diet and susceptibility. Each has an additive effect on incidence. The risk in the population is heavily weighted by the occurrence of the three factors in combination. These situations must be carefully reviewed for interpretation of the data.

Table 6
Multifactorial causal factors

Risk Factors	Subgroup Incidence /100,000	% distrib.	Incidence in pop. from exposure to A
None	1.0	2	
A	1.5	2	1.5 x .02 = .03
B	2.0	2	
C	1.3	2	
AB	3.5 (1.5 + 2.0)	5	3.5 x .05 = .18
BC	3.3	5	
AC	2.8	5	2.8 x .05 = .14
ABC	4.8 (1.5 + 2.0 + 1.3)	77	4.8 x .77 = 3.70

Incidence if only A exposure evaluated = 4.05
Incidence from three factors = 4.28
Incidence in total population based on distribution of subgroups =
4.05 + .02(None) + .04(B) + .026(C) + .165(BC) = 4.30

Summary recommendations

In summary, the challenges in epidemiology in the future will be the examination of weak associations between risk factors and disease. These will usually occur in common diseases in association with common exposures, so they could have a high impact on disease in the population. These associations must be carefully examined for the effect of possible confounding variables since confounders with strong relative risks for that disease and especially those with a marked difference in the distribution of the confounding variable in relation to the exposure of interest could be the sole explanation for the weak association. Weak associations may also be reported which are not real when the investigator has included a large number of variables in a model without carefully considering the hypothesis to be tested and reports significant effects from one or more variables. Each variable with an elevated risk ratio may be reported separately in papers. This method of handling data and publications will prevent the reader from recognising the problems inherent in the study, that is the indiscriminate examination of hundreds of variables in a single study and the separate reporting of each of the multiple weak associations. The papers related to the study do not often discuss the true focus of the original study. The investigators may not report the extent to which they have searched carefully for any potential confounders for each of these associations. These reports of weak associations based on examination of large numbers of variables without prior hypotheses justify the criticism which is often levelled at epidemiologists, that is that they are reporting a new risk every week.

From another viewpoint, epidemiologists may actually be reporting weak associations that are not as weak as observed because of methods of analysis. Of course, they may truly have weak associations because of the nature of the relationship of the risk factor to the etiology of disease. These reports of weak associations may arise because of a growing movement in epidemiology to use available technology to place many variables in models and then look at what falls out as risks rather than carefully examining the interrelationships between factors added to the models, and deciding how they relate to the etiology of disease. Clearly, hypothesis-building should be done in advance of modelling. Correlations between variables should indicate a warning that the relationship may not always be one of a confounding variable but of factors related to the etiology of or biologic pathway to disease. Epidemiologists should be advancing principles which would promote development of different scenarios for disease etiology and test the evidence as well as the sensitivity and uncertainties that surround those scenarios as part of the investigation of weak associations. Most importantly, the investigator must describe these questions and uncertainties regarding the risk estimates and the limits of the methods they used as part of the report. At present, even when the investigator has described the caveats related to the data, the review is not systematic and does not include any numerical comparison which allows evaluation of the certainty of the association except for the usual reliance on statistical tests. The time has come to begin to consider what should be included in reports to allow an evaluation of the certainty which can be placed on a reported weak association. The current procedure of resolving the issue through arguments and multiple reviews is not the best choice for resolution. Further use of statistical tools may help to define and clarify the problems surrounding weak associations.

References

1. Rothman K (1986) Modern Epidemiology. Boston/Toronto: Little Brown And Company
2. Greenland S, Robins J (1985) Confounding and misclassification. AM J Epidemiol; 122:3:495-506
3. Ahlbom A, Steineck G (1992) Aspects of misclassification of confounding factors. AM J Ind Med 21:107-112

Commentary on confounding: Examples of its influence in weak associations

Kunio Aoki, Tokyo / Japan

I would like to introduce an example of confounding and subgroup analysis in relation to interaction of causative factors in a case-control study of oesophageal cancer in Japan. It is an old study, but this analysis is now linked to molecular epidemiology in oesophageal cancer.

Subject and Methods

Oesophageal cancer patients admitted to 3 hospitals in Aichi (low risk area) and 2 hospitals in Wakayama (high risk area) were investigated for the period of 5 years between 1974 and 1979. Two hundred and one cases (145 males and 56 females) and 403 controls matched for age, sex, time of admission and residence area were compared. A major portion of the questionnaire was devoted to dietary habits. Smoking, alcohol drinking, occupational and family histories and other items relevant to oesophageal cancer were also inquired by three experienced interviewers; two for Aichi and one for Wakayama.

Results

Major risk factors associated with oesophageal cancer were smoking, alcohol-drinking, hot tea-gruel-taking habits, salty foods, and low socio-economic conditions. Risk lowering factors were excessive intake of some fruits and vegetables. Relative risks of current smokers were 4.3 in Aichi and 2.1 in Wakayama. Relative risks of every evening sake (Japanese wine) drinkers were 4.0 and 3.7 in males, respectively. In females, risk of smoking was 1.6 and sake drinking was 0.7, although the number of cases were small. The following analysis were done for males only.

Confounding analysis

Confounding effects of the risk factors and risk lowering factors were examined in males (Table 1). Relative risks of those with smoking and sake drinking increased synergistically. Patients in the high risk area (Wakayama) showed much higher effect. When salty foods intake was confounded in these two factors, the risks changed significantly, as shown in Table 1. However, little effect by salty food intake was observed in Aichi (lower risk area), which suggests that some other factors might be interviewed. When cabbage (risk lowering factor) was confounded in place of salty food intake, the risks were remarkably lower, although the size of risks was different by area. Higher risk figures might indicate the small number of cases.

Table 1

Relative risks for oesophageal cancer in combination of risk or protective factors in Aichi (low risk area) and Wakayama (high risk area)

Smoking	Sake drinking in evening	Salty foods	Males in Aichi	Males in Wakayama
		-	4.7**	14.7**
+	+	+	2.6*	10.1**
+	+	-	3.4*	4.5**
-	+	+	1.3	5.6**
-	+	-	2.0	7.0**
-	-	+	0.5	2.6**
-	-	-	1.0	1.0
(Maximum Likelihood Estimate)			1.2	1.3
95% Confidence Interval			0.7-2.0	0.6-3.0

R.R. for salty foods adjusted for smoking and drinking

Smoking	Sake drinking in evening	Cabbage	Males in Aichi	Males in Wakayama
+	+	-	18.4**	77.4**
+	+	+	6.6**	13.7**
+	-	-	14.3**	5.9*
+	-	+	3.4**	5.8**
-	+	-	43.0	13.7**
-	+	+	1.1	5.6**
-	-	-	2.9	4.6
-	-	+	1.0	1.0
(Maximum Likelihood Estimate)			0.3	0.3
95% Confidence Interval			0.1-0.9	0.1-1.1

R.R. for consumption of cabbage adjusted for smoking and drinking

* p<0.05, **p<0.01

Subgroup analysis

Location of primary cancer site in the oesophagus can be classified into three parts, that is, upper, middle and lower regions. We divided cancer cases by location excluding primary site unknown and advanced cases. Table 2 shows relative risks of some diet-related factors by site in both areas combined. Smoking affected more the middle and lower sites and sake drinking less the lower site. Tea-rice gruel (very hot food) intake habit and salty taste favoured, including soy sauce seasoning, showed higher risks only in the lower cancer site. Konjak flour paste and cured fried beans were cheap food at that time, so we took them to be one indicator of poor dietary habits. They showed different ways of risks by site.

Table 2
Foodstuffs with suggestive high risks for esophageal cancer by site (males only)

		Site of Cancer		
		upper	middle	lower
Smoking	yes/no	2.8	4.4	4.3
Sake drinking in evening	yes/no	2.9	2.7	-
Tea-rice gruel	yes/no	-	-	3.3
Salty taste favored	yes/no	-	-	2.1
Vegetables seasonsed by soy-sauce	yes/no	-	-	3.3
Konjak flour paste (Konnyaku) ever used/ no use		2.2	7.3	-
Fried bean curd (Aburage) ever used/no use		7.4	3.9	3.0

Table 3 shows risk-reducing effects for fruits and vegetables by site. Cabbage, tomato and carrot reduced risk remarkably on the upper cancer site. Strawberry and watermelon showed less effect on the upper site. Cooked beans also remarkably reduced risk on the upper site.

Table 3
Foodstuffs with suggestive low risks for oesophageal cancer by site (males only)

		Relative risk by site		
		upper	middle	lower
Cabbage	use/no use	0.1	0.3	0.3
Tomato	use/no use	0.1	0.5	-
Carrot	use/no use	0.2	0.6	0.3
Strawberry	use/no use	0.5	0.3	-
Watermelon	use/no use	0.4	0.3	0.5
Cooked beans	use/no use	0.1	0.6	0.5

Discussion

Not so many cases might cause fluctuated figures of risk, but we could obtain tendencies of risks by causative factors. Confounding of smoking and alcohol drinking is wellknown. This study suggests the effects of other factors. The narrow oesophageal tube can be divided into three sections: upper, middle and lower (oesophagus and stomach junction), where much stimuli by foods and drinks were loaded. Speed of food swallowed is faster until it reaches the middle section and slows down in the lower section. These anatomical and physiological findings might be associated with cancer incidence. This study seems to endorse partly such a hypothesis. In most countries, declining trend in oesophageal cancer appeared first in the upper, then the middle and lastly the lower sections. Risk lowering factors may suggest some roles in the reduction of this cancer. High risk area was economically poor and Drs Kammon and Hirayama reported that blacken fern was a risk factor of oesophageal cancer, but this could not be confirmed. It may be due to a difference in study subjects. Recently, a molecular epidemiological study was started to compare the findings between cancers in the three sections of the oesophagus, because it was difficult to compare with those of normal people.

References

1. Aoki K, Okada H, Segi M. et al. (1980) Causative factors of oesophageal cancer investigation by case-control study. Nagoyo T, Tominaga S. eds. Cancer-Japan and the World - The trends and etiology, pp.300-315, Shinohara Shuppan, (in Japanese)
2. Aoki K, Sasaki R, Okada K, Haenszel W (1986) Epidemiological study on esophageal cancer in Japan. Kasai, M., ed., Esophageal Cancer, pp. 3-6, Excerpta Medica, Amsterdam
3. Kammon S, Hirayama T (1980) Study on epidemiologicic features and etiology of esophageal cancer. Nagoya T, Tominaga S, eds. Cancer - Japan and the World. Its trends and etiology, pp 279-299, Shinohara Shuppan, Tokyo (in Japanese)

Bias in observational studies·

Hans Hoffmeister, Berlin / Germany

1 Bias as a major problem in modern epidemiology

Before presenting more specific remarks on examples of observation biases and proposals on how to avoid or minimise such biases, I would like to address some issues concerning biases in general.

1. No generally accepted and used definitions of biases are to be found in textbooks and in the scientific literature. In the Oxford University Press "Dictionary of Epidemiology" of 1983, bias is defined as "any effect at any stage of investigation or inference tending to produce results that depart systematically from the true value". This definition is clear, but comprehensive. It would include all the possible main categories of bias, namely selection bias (resulting from non-representativeness of the population studied); observation bias (resulting from errors in determining exposure or disease status precisely), and uncontrolled confounding and even unprecision of measurement (resulting from failure to fully account for effects of other risk factors). It seems appropriate to me to take "bias" as a generic term for all sources of systematic distortions of results which lowers their accuracy, although simple measurement error and confounding are often not included in discussions on biases.

2. Several proposals have been made for sub-classification of biases, but no terminology has been generally accepted. It seems to me that there is an urgent need to have a discussion within the scientific community aimed at establishing and publishing a catalogue of biases, in which each type of bias is clearly characterised and given a specific name. In addition, there should be a clear description of which situations specific biases cover or do not cover. One example is the possible confusion in the literature by using the different terms such as observation bias, information bias and misclassification bias to describe what appears to be the same thing. Do all of us understand the three expressions really as synonyms? Hereafter I will keep to the term "information bias".

3. An attempt should be made not only to establish defined terms but also to define rules, checks and procedures to avoid or minimise a specific bias or at least to control it. There should be a generally known and accepted checklist for researchers on the planning or conducting of a study which contains a description and possible sources of bias as well as advice on how to take them into account. It

does not seem appropriate to criticise studies as being biased after they have been published - considerations of bias should be taken into account before publication.

4. Where the extent of a particular source of bias cannot be accurately quantified, authors should include results of sensitivity analysis demonstrating the extent to which their main findings are dependent on this source of bias, if possible. It is simply not enough to refer to the existence of a possible bias in the discussion section of a paper without giving some information on its likely importance. This means that epidemiologists should become more aware of procedures that have been developed to correct for bias, and that more work should be undertaken to develop such procedures further. Also, there is an increasing need to collect information on validity and reliability, so that these correction procedures can be properly employed. Correction for bias is particularly important when studying weak associations, else the "signal" be blotted out by the "noise".

2 Subclassification of biases

Some interesting proposals, which might serve as a basis for something like a "standard classification of biases" already exist. One noteworthy definition was published by Steineck and Ahlbom (1992). I would like to point out the following: compared to other attempts at biases classification, Ahlbom's concept contains a new and very important step, namely it attempts to set up rules quantifying attributable subclasses of biases and calculate from that a total bias value.

Sackett (1979) published a catalogue of more than fifty different forms of biases using a classification according to the stages of research in which a bias can occur. Half of the individual types of biases listed can be characterised as information biases. On closer inspection, many of these can also be classified as selection biases so that the exposed and non-exposed or the diseased and non-diseased will be selected in a different way. In other words, the primary cause of a selection bias might be a differential observation. Information and selection bias are not necessarily independent.

A more recent classification of biases was published by Choi and Noseworthy (1992). These authors used a classification and sub-classification not only according to the type of bias but also to the type of study. Figure 1 presents a modification of this classification for information biases. I have added the items "desirability bias", "knowledge bias" and "social strata bias" which, from our experience, are especially important when weak associations are to be detected.

It is helpful to characterise biases as to whether, for a particular study type, they are likely to cause over- or under-estimation of the association of interest or whether they will have no effect. As indicated by the larger arrows in Figure 1 information biases often result in over-estimation of risks. However, depending on the observer's and the subject's attitudes and beliefs, underestimation of risk can also occur. In cohort studies some of the biases do not exist.

Figure 1

Information Bias

Type of Bias	Type of Study			
	Cross-sectional	Case-Control	Retrospective Cohort	Prospective Cohort
Recall Bias	↑ ↓	↑ ↓	NP	NP
Knowledge Bias	↑ ↓	↑ ↓	NP	NP
Questionnaire Bias	↑ ↓	↑ ↓	↑ ↓	↑ ↓
Desirability Bias	↑ ↓	↑ ↓	NP	NP
Social strata Bias	↑ ↓	↑ ↓	↓ ↑	↓ ↑
Diagnostic Suspicion Bias	↑ ↓	↑ ↓	↑ ↓	↑ ↓
Exposure Suspicion Bias	↑ ↓	↑ ↓	NP	NP
Intra-Interviewer Bias	↑ ↓	↑ ↓	↑ ↓	↑ ↓
Inter- Interviewer Bias	↑ ↓	↑ ↓	↑ ↓	↑ ↓

NP = Not present

Source: B.C.K. Choi and A.L. Noseworthy, 1992 (modified).

In my other contribution, we restricted ourselves to the first five types of information biases listed in Figure 1. They mainly refer to the study subjects, whereas the last four refer to the observers. According to our experience, difficulties in interpreting studies dealing with the menace of daily life, especially the many contradictory study results on environmental agents and cancer, mainly arising from biases based on differential recall, knowledge, attitudes, desirability and socio-economic status in the study populations being compared.

3 Examples of information biases

Recall bias is a frequent handicap in cross-sectional and case-control studies. If long latency periods exist between an exposure and the onset of disease, differential recall bias is quite likely to play an important role. In case-control studies, especially those using general population controls, the cases will often be more likely to recall an exposure than the controls because of their search for an explanation of their disease.

"Desirability bias" seems to be particularly important when small effects of environmental agents are to be detected. It is often the case that the agent is already considered by the public as a cause of disease, before a study is done. This happens, for example, with dioxin, electromagnetic fields, summer smog, and many other environmental hazards. If cases overestimate exposure because they believe it causes their disease, an association is bound to be found, regardless of whether or not an association exists - a self-fulfilling prophecy.

Another type of bias in this subclassification that has been included into the list of Choi and Noseworthy (1992) is the social strata bias. This bias does not mean those real differences in exposure or disease frequency in different socio-economic groups, which can be considered to be confounders. An additional distortion can occur by unequal distribution of socio-economic classes in the study population. This distortion originates in different knowledge, attitudes and believes within the different social classes. I will give a trivial example to demonstrate what is meant by this:

In the German National Health Survey, the participants were asked for their body weight, and this parameter was also precisely measured. Figure 2 shows, in the form of regression lines, the relationship between the subjective statements and measured values. The figure shows clearly that the more obese a person is, the more the body weight is underestimated. Women tend to underestimate their weight more than men. In every German study population this bias would lead to an underestimation of the risk from body mass index even when the bias is distributed equally between the groups being compared. Namely, a non-differential bias would occur. This bias can be corrected for by using the regression equation shown in Figure 2.

Figure 2

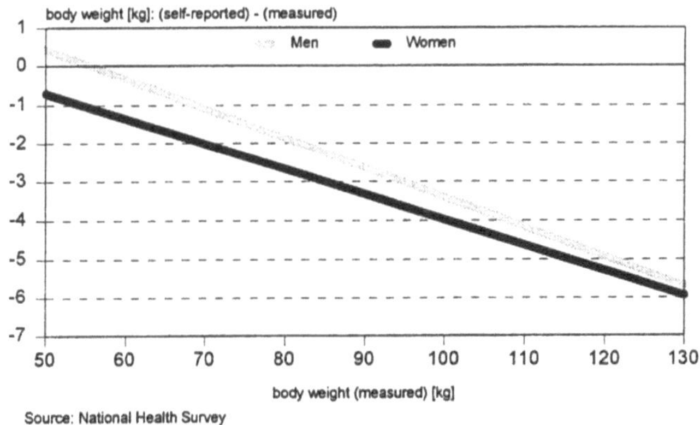

It is of greater importance, however, when the exposed and non-exposed or the diseased and non-diseased differ with respect to social strata, which is often the case. The magnitude of this additional social strata bias can be seen in Figure 3.

Figure 3

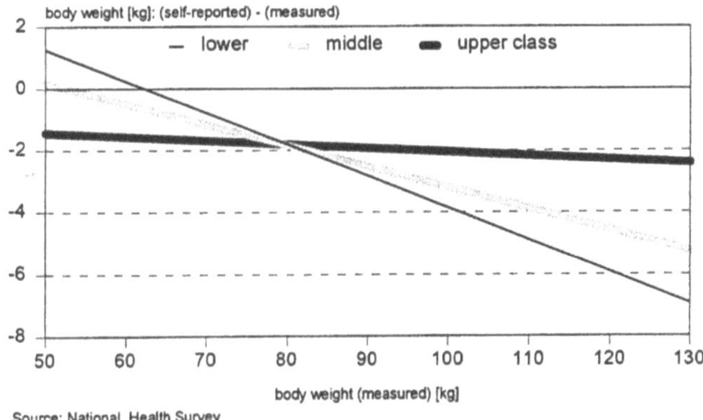

Upper class persons report, independently on their own body weight, that their body weight is only 1 kg less than it actually is. People in the lowest social class, however, underestimate their body weight by up to 6 kg in the most obese group.

4 Methods of avoiding or minimising information biases

It is possible in theory to measure quantitatively the impact of particular biases by using standardised questions or scales, evaluated earlier in a rigid way. In addition, recall biases, desirability biases, knowledge biases and questionnaire biases can be reduced by giving careful help to study participants (by using structured interviews or questionnaires) and reminding all participants (including the controls) of situations, where typical exposures might have occured.

Another point I would like to make is that the questionnaire or details of a structured interview are rarely described in publications. However, for comparing contradicting study results, this information is indispensable. It should be a standard condition when accepting papers for publication, that authors make available to the reviewers full details of all procedures used in measuring exposure, disease and even confounders where it is not practical, for reasons of space, to include full details in the paper itself.

Using biomarkers in a subsample of the study population is a possibility to be considered in the evaluation of questions or scales. Wherever possible, a measurement should either support interviews or questionnaires or replace them completely. This can be an effective way of avoiding many of the information biases. Biomarkers, rejecting long-term exposure to certain substances can be measured for example in body fat or in hair, nails and many other parts, thus giving more accurate data on exposure than those obtained from interviews. In several ecological and nutritional studies, biomarkers of exposure, such as pesticides, dioxin, and heavy metals have been successfully used. Other chemical elements (like selenium) have been determined as well as markers for specific nutritional habits, such as vitamin E, carotenoids and linoleic acid. Often, the determination of diagnostic parameters in blood samples will not only be the most accurate but it is also the cheapest method of testing for a possible disease or risk. One example of the advantages of using biomarkers relates to drug consumption. Even when the consumption of a drug is carefully investigated, not only by asking study participants about their intake and compliance but also by listing the drugs they showed to the investigator, analysing the level of the drug in serum often gives more precise information on it. If blood is already taken in the study, this method might not be more expensive than questioning.

In a study conducted by us, only 32 % of those persons reporting regular intake of digoxin as prescribed by their doctors actually had serum levels of the active substance consistent with this statement. This demonstrates a large desirability bias regarding the use of this chronic medication. Serum levels of many broadly used drugs like caffeine, salicilic acid, theophylline and benzodiazepines can be easily determined (Melchert & Hoffmeister 1994; Melchert & Hoffmeister 1994; Melchert & Hoffmeister 1994).

It should be realised that biomarkers are not always superior to questions. Chemicals with a short half-life can only be used to determine recent exposures. Even those with a much longer half-life may be inadequate to detect risks resulting from exposure early in life.

Figure 4a shows the magnitude of a bias when comparing answers to questions on hypertension as compared to that measured, as a reference. This type of bias arises partly from failure of memory and partly from lack of knowledge. The number of false negative answers (twice the number as correct answers) was so large that even a slightly unequal distribution of study participants between exposed and not-exposed would lead to an erroneous relative risk. In particular a different distribution of the socio-economic groups between the comparison groups would lead to a biased result. As can be seen from Figure 4b, in Germany, the upper class is much better informed about their blood pressure than the rest of the population.

Figure 4a
Reported and measured hypertension
West-Germany, men, 40-55 years, n=924

| | self-reported | |
measured	yes	no
yes	10,5%	19,5%
no	3,0%	67,0%

Source: National Health Survey 1991

Figure 4b
Reported and measured hypertension by social class
West-Germany, men, 40-55 years, n=924

social class	reporting false negative	reporting false positive
upper	13,7%	2,6%
middle	21,4%	3,8%
lower	21,9%	1,3%

Source: National Health Survey 1991

Figures 5a and 5b show false-negative and false positive rates of self-reported allergies of the respiratory tract, using as reference a serum test of sensitisation against the total of respiratoric allergenes (Bellach 1992). The reporting bias in both directions is only around 10 percent. A differential information bias by social class is again evident.

Figure 5a
Reported and measured allergy
West-Germany, women, 40-55 years, n=907

measured	self-reported	
	yes	no
yes	6,6%	10,9%
no	9,0%	73,5%

Source: National Health Survey 1991

Figure 5b
Reported and measured allergy by social strata
West-Germany, women, 40-55 years, n=907

social class	reporting false negative	reporting false positive
upper	16,8%	10,3%
middle	9,9%	8,9%
lower	6,2%	7,9%

Source: National Health Survey 1991

While analysing National Survey data in the East and West German population, an interesting phenomenon was observed. Those participants aged 40 years and older, who lived in an undivided Germany during their childhood, showed no differences in the reported prevalence of respiratoric allergies. However, the younger the West German study population, the greater the number of persons stated to be suffering from an allergy. This was not the case in the East German population. We would never have believed this result, knowing how deeply concerned the younger West Germans are about the poisened environment that is continually endangering their health. However, we could completely confirm the reported prevalences in the younger West and East Germans using the sensitisation test. Due to this "biomarker" we now know that a steady increase in respiratoric allergies really occurred in West Germany and that the cause for this trend can be found in childhood.

When the biases demonstrated before are equally distributed in the exposed and non-exposed, there remains a problem of non-differential bias: Non-differential errors in information collected on exposure and disease influence the magnitude or intensity of the observed or measured risk, resulting in a dilution effect. This dilution bias always leads to an underestimation of the risk. Small true risks can easily disappear in the noise of non-differential errors.

It should be realised, however, that non-differential errors in determining confounding variables do not necessarily lead to a dilution effect. Such errors may result in failing to fully control for confounding, thus leading to an overestimation of the true relationship.

5 Social unequalities as a source of biases

Well known confounders such as smoking, alcohol consumption, specific nutritional habits and other life style variables which may have a real impact on the disease in question can easily be influenced by information bias, thus causing severe problems. Persons exposed to an environmental agent may differ in their attitudes to a confounder from persons who are not exposed. Persons having a disease strongly influenced by one of the confounders may report their own behaviour, with respect to this confounder, differently from non-diseased persons. Confounders therefore have to be controlled for information biases as well.

In a cross sectional study we ranked people by their anxiety concerning environmental health hazards. (Figure 6). Smokers were reported to be remarkably tensed more often than non-smokers. This anxiety may lead to smokers being more likely to report exposure to environmental agents, and so confuse analysis of the joint effect of smoking and environmental agents on certain diseases.

Figure 6

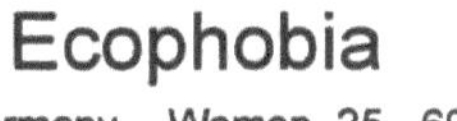

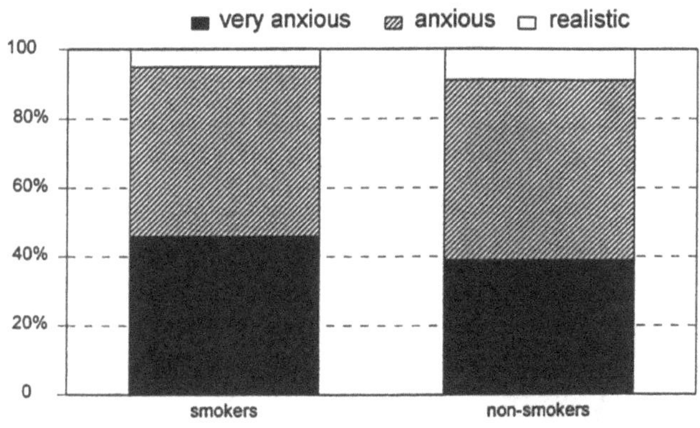

In many epidemiological studies of weak associations, distortion caused by social inhomogenity may be the leading source of bias as well as of confounding. Avoidance or better control of differences in socio-economic status would be the most effective way of getting more accurate and reproducible results for studies where weak associations or even non-associations have to be ascertained.

Control for socio-economic disparities might usefully be based on an index which takes into account education, occupational status and income. This "integrative confounder" should be added to those confounders normally controlled for, such as smoking and alcohol consumption. Controlling for social strata will also partly control for other variables with which it is correlated, such as nutrition, physical activity and other lifestyle characteristics.

Figure 7 shows the impact of controlling for the confounder "social strata" and thus for all the biases with it.

Figure 7
High Fat Intake and Hypertension
West-Germany, men, 25-69 years

variable	model 1		model 2		model 3	
	RR	p	RR	p	RR	p
fat intake (100 kcal/day)	1.10	ns	1.13	0.03	1.35	<0.001
energy intake (100 kcal/day)	0.96	ns	0.94	0.04	0.85	<0.001
age (years)	1.04	<0.001	1.04	<0.001	1.04	<0.001
social class: lower	—	—	1.00	—	1.00	—
middle	—	—	0.72	ns	0.67	ns
upper	—	—	0.50	0.01	0.52	0.04
smoker (yes/no)	—	—	—	—	0.82	ns
alcohol consumption (10 g/day)	—	—	—	—	1.24	<0.001
body mass index (1 kg/m^2)	—	—	—	—	1.24	<0.001

Source: National Health Survey 1991

We used three models to test for a hypothesised small effect of fat intake on high blood pressure. The data were derived from a cross-sectional study, where blood pressure was measured precisely and nutrition was investigated by diet history interviews. In a first model, total energy intake and age were controlled for. No significant effect of fat intake could be detected. In the second model, social strata was introduced as an additional variable. A small, but statically significant influence of fat intake could now be seen. In a third model, controlling for some other important confounders, the effect became even more visible. A model without the social strata variables showed a weaker association, suggesting that well-known confounders and biases like body mass index and alcohol consumption are partly considered by a model that includes the social strata.

References

1. Steineck G, Ahlbom A (1992) A definition of bias founded on the concept of the study base. Epidemiology 3: 477-482
2. Sackett DL (1979) Bias in analytical research. J Chron Dis 32: 51-63
3. Choi BCK, Noseworthy AL (1992) Classification direction and prevention of bias in epidemiologic research. JOM 34: 265-271
4. Melchert HU, Hoffmeister H (1994) Representative consumption data of caffeine-containing drugs and serum levels of caffeine of drug users and non-users in Germany. Pharma epidemiol drug safety 3 suppl 1,3
5. Melchert HU, Hoffmeister H (1994) Representative consumption of salicylic acid-containing drugs and serum levels of salicylic acid user in Germany. Pharma epidemiol drug safety 3 suppl 1,3
6. Bellach BM et al. (1992) Inhalation allergies (in German). In: Health of the Germans. Inst Social Med and Epidem, Fed Health Office Berlin. Ed Hoffmeister H

Research strategies for assessing epidemiolgic associations, in relation to the distribution and measurement of exposures

Ross L. Prentice, Seattle / USA

1. Introductory remarks

It seems important to distinguish epidemiologic effects that are small because the exposure (or characteristic) is unimportant for the study disease from effects that are small by virtue of study design and execution. In the former situation there is little variation in disease risk across the exposure levels that are within the range of common human experience and it follows that only a small fraction of current human disease burden can be attributed to the particular exposure. In the latter situation, a careful scrutiny of potential study populations, study designs, and exposure assessment instruments is needed to optimise the reliability and efficiency of research to assess the exposure-disease association.

This brief outline focuses on the latter situation. An exposure, or set of exposures, is hypothesised to have an important relationship to a disease within the range of common human experience. If the exposure distribution within a convenient study population is broad enough so that the within-population risk gradient is still fairly large, and if the exposure and confounding factors are well measured, a cohort or case-control study design may well provide an efficient vehicle for assessing exposure effects on the disease. In some such situations a randomised controlled intervention trial may be merited if the public health implications are great, since the randomised trial can also control for unmeasured (and poorly measured) confounding factors, and because the randomised trial can directly assess the health impact of a highly specialised exposure change, for which sufficient observational data may never be available. However, if the exposure distribution within populations conveniently available for study is narrow, or if the exposure or confounding factors are poorly measured, the reliability of cohort and case-control studies is typically much reduced, and alternative research strategies may be needed.

Concerning the breadth of the exposure distribution, one can consider an extreme situation in which the individual exposures are identical within each available study population. A cohort or case-control study within any such population, or a meta-analysis of such studies involving multiple populations would be uninformative concerning the exposure-disease association as all pertinent information resides between rather than within populations. An aggregate data (ecologic) study is then clearly the only viable observational research strategy. As a less extreme situation, suppose that the breadth of the exposure distribution within populations is a fraction of that across populations. For example, suppose that the within population exposure variance is about 25% of that across populations. What then is the preferred research strategy? The answer very likely depends on a variety of factors including knowledge about within-population and between-population confounding factors, and the ability to accurately measure both the exposures under study and confounding factors, and comparative research costs.

The measurement error issue is particularly important in epidemiology, but our ability to anticipate the effects of such error, and to properly allow for measurement error in data analysis, is still quite limited. Data that are typically available (e.g. reliability subsample) allow only the simplest of measurement models to be applied. Most reports in the epidemiologic literature, after some attempt to "validate" the exposure assessment instrument, totally ignore measurement error in presenting estimates of epidemiologic effects (e.g. relative risk estimates). On the other hand, it is clear that the particular properties of a measurement instrument may have profound implications for the ability to identify and quantify epidemiologic associations, especially if the exposure distribution is fairly narrow in the study population.

Aggregate data research strategies appear to have considerable potential for reducing the impact of measurement error in exposure and confounding factors. Such studies may be designed and analysed in a manner that uses only exposure and confounding factor information that has been averaged over, say, a few hundred individuals, thereby essentially eliminating the noise aspect of measurement error. These facts suggest that aggregate data studies merit a greater role in the assessment of the epidemiologic effects of poorly measured exposures, though the ability to adequately control for (between-population) confounding factors is an issue in such analysis.

In the next section the potential impact of limited exposure range and of exposure measurement on the assessment of epidemiologic effects will be illustrated by examining available data on the relationship of dietary fat to breast cancer. The subsequent section will describe some other research strategies in the diet and cancer context.

2. Analytic epidemiologic study of dietary fat and postmenopausal breast cancer

The hypothesis that a low fat diet may reduce the risk of breast cancer has been propagated for several decades. Experimental studies in rodents indicate specific roles for both fat reduction and calorie restriction in inhibiting mammary tumorigenesis (Tannenbaum 1942; Newberne at al. 1989; Freedman et al. 1990). International correlational studies suggest strong relationships between breast cancer incidence and mortality and fat consumption, particularly for postmenopausal women (Carroll et al. 1995; Prentice & Sheppard 1990); and they gain support from time trend and migrant studies. For example, we used breast cancer incidence rates among women in the age range of 55-69 years in 21 countries with representative cancer registration and per capita nutrient supply data to suggest that a 50% reduction in total fat intake in the United States could eventually reduce postmenopausal breast cancer incidence rates to about 40% of present levels (Prentice & Sheppard 1990). There have been a large number of case-control studies of this association reported over the past two decades. Howe et al. (1990) carried out a joint analysis of the data from 12 such case-control studies that included 4247 breast cancer patients and 6095 control subjects, about two-thirds of whom were postmenopausal. The authors reported a highly significant (p<.0001) positive association between breast cancer risk and estimated total fat intake among postmenopausal women, with estimated relative risks of 1, 1.20, 1.24, and 1.46 across fat intake quintiles. However, this trend was interpreted as being less than would be anticipated from the international correlational analysis. Recently, Hunter el. al. (1996) reported on a pooled analysis of seven cohort studies of dietary fat an breast cancer, which included 4980 breast cancer cases arising in the follow-up of over 300,000 women. They reported an estimated relative risk of only 1.05 (95% confidence interval 0.94 to 1.16) for the highest as compared to the lowest quintile of calorie-adjusted total fat intake. The authors noted a similar lack of trend across percent energy from fat quintiles, with somewhat elevated breast cancer incidence among women whose reported energy intake from fat was less than 15% or 20%. The seven cohort studies each used a food frequency instrument for dietary assessment and included a "validation" subsample in which an additional more comprehensive dietary approach, involving multiple food records or recalls, was used. Hunter and colleagues (1996) commented that using these validation data to provide a measurement error correction had little impact on their analysis. Differences between these cohort and case-control study results could be due to differing populations studied and differing dietary instruments, to differing control for energy intake, to recall bias in the case-control studies or, importantly, to dietary measurement error biases in one or both sets of studies.

Existing measurement error models (Rosner et al. 1989; Rosner et al. 1990; Wacholder et al. 1993; Carroll et al. 1995) assume either that the validation sample data are without measurement error or that any such error is independent of both the true dietary exposure and all other study subject characteristics. However, measurement errors for protein consumption under various dietary self-report instruments have been shown to be correlated (Plummer & Clayton 1993). Also there is a growing body of literature (Lichtman et al. 1992; Bandini et al. 1990; Heitman & Lessner 1995; Martin et al. 1996; Sawaya et al. 1996) using doubly labelled water and other techniques to obtain objective measures of total energy expenditure while controlling for physical activity. These studies consistently show obese persons to substantially underreport calorie intake, perhaps by

25-50%, on food records and other forms of dietary self-report. For example, a recent study in Denmark (Heitman & Lessner 1995) indicated that self-reported energy intake among middle-aged and older women was underestimated in an approximate linear fashion across deciles of percent body fat. The estimated degree of underreporting increased from near zero at the lowest body fat decile to 30-40% in the upper three deciles. Furthermore, protein energy calculated from urinary nitrogen output, was underreported to a considerably lesser degree than was total energy, making it likely that both total fat and percent energy from fat are increasingly underreported as percent body fat increases.

Prentice (1990) considered a more relaxed measurement model in this context by allowing all measurement error parameters to depend on body mass index (BMI) (i.e. weight in kg divided by square of height in meters) categories and by incorporating a random underreporting quantity that applies to each dietary self-report instrument (1996). The implication of this measurement model improvement was then examined towards an understanding of the varying results of epidemiologic studies of dietary fat and breast cancer among postmenopausal women.

The related measurement model alluded to above was applied to food record and food frequency data from the NCI-sponsored Women's Health Trial (Insull et al. 1990; Henderson et al. 1990). The WHT was a randomised and controlled feasibility trial of a low-fat dietary pattern, carried out among 303 women (184 intervention, 119 control) in the age range of 45-50 years that took place from 1985 through 1986 in Cincinnati, Houston and Seattle. Specifically, four-day food record (4DFR) and food frequency questionnaire (FFQ) data (Willett et al. 1985), along with height and weight data at baseline on all WHT women were used in conjunction with 4DFR and FFQ data (Block et al. 1986) one year later on dietary control women to build a measurement model for both total (daily) fat intake and for percent energy from fat.

To characterise the influence of measurement error in fat intake assessment on the results of analytic epidemiologic studies, the strong associations between measures of dietary fat and postmenopausal breast cancer seen in international correlational analysis (Prentice & Sheppard 1990) was supposed to be due entirely to fat consumption. The measurement model developed using Women's Health Trial data was then used to examine the extent to which such strong associations are attenuated and distorted by measurement error in dietary assessment. For example, if the strong signal assumed from international comparisons is largely masked by measurement error in 4DFRs and FFQs, it would follow that these instruments may well be inadequate for the reliable assessment of trends between disease risk and fat intake in cohort or case-control studies, regardless of study size.

Table 1 shows some of the results from these exercises. If measurement errors were entirely overlooked, the international data analysis can be projected to imply relative risk trends from 1 to 3.08 across 4DFR quintiles of total fat, and from 1 to

4.00 across FFQ quintiles of total fat, with quintiles determined by baseline data in the WHT. Similarly, one can project relative risk trends from 1 to about 1.8 across WHT quintiles of percent energy from fat with either the 4DFR or the FFQ. If a classical measurement model is assumed so that provision is made for the noise aspect of dietary fat measurement but not for any systematic underreporting as a function of body mass, these relative risks are substantially attenuated towards unity, with relative risks ranging from 1 to 1.54 (4DFR) and from 1 to 1.27 (FFQ) across total fat quintiles, and from 1 to 1.28 (4DFR) and from 1.27 (FFQ) across percent energy from fat quintiles. Furthermore, as shown in the final row of each section of Table 1, these projected relative risks are reduced much further by allowing an underreporting bias (i.e., systematic bias) for each individual with distribution that varies as a function of body mass. Specifically, when both measurement noise and systematic biases are accommodated, the projected relative risks across WHT fat intake quintiles range only from 1 to about 1.1 for total fat and percent energy from fat with either assessment instrument.

Table 1
International data projected relative risks of breast cancer at various percentiles of the Women's Health Trial fat intake distribution, under various measurement error models

Measurement Error	4DFR Quintile					FFQ Quintile				
Overlooked	1 †	2	3	4	5	1 †	2	3	4	5
Log-Total Fat										
Systematic and Noise	1	1.39	1.75	2.21	3.08	1	1.51	2.00	2.66	4.00
Systematic	1	1.13	1.24	1.36	1.54	1	1.11	1.19	1.28	1.42
Neither	1	1.01	1.03	1.05	1.09	1	1.00	1.02	1.05	1.11
% Energy for fat										
Systematic and Noise	1	1.20	1.35	1.53	1.82	1	1.20	1.36	1.55	1.86
Systematic	1	1.07	1.13	1.20	1.28	1	1.07	1.12	1.19	1.27
Neither	1	1.00	1.02	1.05	1.11	1	1.02	1.04	1.06	1.11

†reference category

It is interesting to note that the measurement model that accommodates both systematic and noise aspects of measurement error yields relative risks in excess of unity at very low levels of percent energy from fat. For example, projected relative risks at 15% versus 30% energy from fat are about 1.06 for the 4DFR and 1.18 for the FFQ. This occurs because such extreme values are predicted to arise from women having elevated body mass whose actual, but not reported, fat intakes tend to be more than 30% of calories.

These analyses suggest that the analytic and economic sources of data on the issue of dietary fat and postmenopausal breast cancer may not be discrepant. Rather, the type of exposure assessment instruments used in cohort and case-control studies may yield fat intake estimates that are so noisy that even a moderate systematic bias can overwhelm the relative risk estimation in an analytic study, yielding results that are highly unreliable. Control for body mass index in such a situation may only make the situation worse as such index may be a more reliable measure of fat intake than is the fat intake assessment, so that its incorporation in the analysis may constitute over-control and give rise to relative risk trend reversals and other anomalies.

3. Other research strategies for the study of dietary fat and disease risk

The measurement model mentioned above suggests that 4DFR's and FFQ's are very weak instruments for use in analytic epidemiologic studies of dietary fat and disease. Only a small fraction of the reported variation in dietary fat is attributable to true variations in fat intake, and this latter variation may be dominated by systematic and random measurement errors. Hence other research strategies may be necessary to make progress in this research area. One such strategy is that of full-scale dietary intervention trials with disease outcomes, as is included in the Women's Health Initiative (Rossouw et al. 1995), though cost and logistics imply that very few intervention trials of this type will be practical. Another promising strategy would improve upon international correlational analysis by relating age- and sex-specific disease rates from good quality cancer registries world-wide to corresponding dietary recall and cancer risk factor data obtained by surveying moderate numbers of study subjects in the catchment areas of such registries (Prentice & Sheppard 1995; Sheppard & Prentice 1995). By virtue of aggregating dietary and confounding factor data among study subjects in a given registry area, one can essentially eliminate the noise aspect of measurement error, while the potential wide variation among dietary habits of populations covered by existing cancer registries further reduces the systematic error-to-signal ratio in dietary self-report. A study of this type is currently in the planning stages. Design exercises to date (Sheppard et al. 1996) indicate that most of the between-population information on dietary fat and breast cancer can be obtained with as few as 199 surveyed women in each 10-year age group in a given population, and that 30 or more populations may be sufficient for reliable and precise relative risk estimation even if several control variables are included in the analysis. Limitations of this methodology relate primarily to uncertainty about the ability to adequately control between-population confounding. For example, the aggregate data study of diet and cancer being planned would include a cancer risk factor questionnaire and blood specimen collecting among surveyed instruments along with a 24-hour dietary recall and a 24-hour physical activity recall. Such recall data may be altogether too noisy for use in a cohort or case-control study as these designs rely on the accuracy of the exposure assessment of individuals for their validity, whereas such noise is expected to have minimal effect on the aggregate data results.

4. Discussion and conclusion

The last two sections focused on dietary fat in relation to breast cancer. The issues of exposure range and assessment apply equally well to many other diet and disease associations, as well as to disease associations with other difficult to measure "exposures", including physical activity and energy balance. Analytic epidemiologic studies of such important associations would be much strengthened if objective measures of exposures could be obtained on a subsample of adequate size, even if such objective measures were rather noisy. For example, biomarker measures of macronutrient exposures may be practical at present and could be used to limit the effects of systematic bias in macronutrient intake self-report. The use of such objective measures and demonstrable consistency of relative risk estimates between analytic and aggregate data studies could provide a rather compelling epidemiologic effects assessment. The use of randomised controlled trials could be restricted to the interventions having the greatest public health potential, or to those where a confluence of results from various observational study approaches is not achieved.

Epidemiologic effects may appear to be small, even when the exposure or characteristic under study has an important role in disease causation. For example, such small effects may arise because of the selection of a study population having limited variation in exposure, or because the exposure is poorly measured. In these circumstances, cohort or case-control studies can only be expected to yield reliable results if the measurement properties of the exposure assessment instrument are carefully estimated and accommodated in data analysis, and if there is an accurate and thorough control for confounding factors. Aggregate data (ecologic) studies, if properly conducted, may yield more reliable assessments of epidemiologic effects because of a broader exposure distribution, and because the noise aspect of measurement error can be essentially eliminated by aggregating over individuals. These points have been illustrated by reference to the controversial issue of dietary fat in relation to the risk of breast cancer.

Acknowledgements

The author would like to thank Joy Poutre for technical support in the manuscript preparation. This work was supported by NCI grant CA-53996.

References

1. Tannenbaum A (1942) Genesis and growth of tumors. III. Effects of a high fat diet. Cancer Res. 2: 468-475

2. Newberne PM, Shrager TE, Conner MW (1989) Experimental evidence on the nutritional prevention of cancer. In: Moon TE, Micozzi MS (eds) Nutrition and cancer prevention: Investigating the role of micronutrients. New York: Marcel Dekker: 33-82
3. Freedman L, Clifford C, Messina MW (1990) Analysis of dietary fat, calories, body weight, and the development of mammary tumors in rats and mice: A review. Cancer Res 50: 5710-5719
4. Carroll RJ, Ruppert D, Stefanski LA (1995) Measurement error in nonlinear models. Chapman and Hall, New York
5. Prentice RL, Sheppard L (1990) Dietary fat and cancer: consistency of the epidemiologic data, and disease prevention that may follow from a practical reduction in fat consumption. Cancer Causes and Control 1: 81-97
6. Howe GR, Hirohata T, Hislop G, Iscovich JM, Yuan JM, Katsonyanni K, Lubin F, Marubini E, Modan B, Rohan T, Toniolo P, Shunzhang Y (1990) Dietary factors and risk of breast cancer: combined analysis of 12 case-control studies. J Natl Cancer Inst 82: 561-569
7. Hunter DJ, Spiegelman D, Adami HO, Beeson L, van den Brandt PA, Folsom AR, Fraser GE, Goldbohm A, Graham S, Howe GR, Kushi LH, Marshall JR, MdDermott A, Miller AB, Spiezer FE, Wolk A, Yuan SS, Willett W (1996) Cohort studies of fat intake and the risk of breast cancer - a pooled analysis. New Engl J Med 334: 356-361
8. Rosner B, Willett WC, Spiegelman D (1989) Correction of logistic regression relative risk estimates and confidence intervals for systematic within-person measurement error. Stat Med 8: 1051-1069
9. Rosner B, Spiegelman D, Willett WC (1990) Correction of logistic regression relative risk estimates and confidence intervals for measurement error: the case of multiple covariates measured with error. Am J Epidemiol 132: 734-745
10. Wacholder S, Armstrong B, Hartge P (1993) Validation studies using an alloyed gold standard. Am J Epidemiol 137: 1251-1258
11. Carroll RJ, Ruppert D, Stefanski LA (1995) Measurement error in nonlinear models. Chapman and Hall, New York
12. Plummer M, Clayton D (1993) Measurement error in dietary assessment: an assessment using covariance structured models. Part II Stat Med 12: 937-948
13. Lichtman SW, Pisarska K, Berman ER, Pestone M, Dowling H, Offenbacker E, Weisel H, Heshka S, Matthews DE, Heymsfield SB (1992) Discrepancy between self-reported and actual calorie intake and exercise in obese subjects. N Engl J Med 327: 1893-1898
14. Bandini LG, Schoeller DA, Cyr HN, Dietz WH (1990) Validity of reported energy intake in obese and non-obese adolescents. Am J Clin Nutr 52: 421-425
15. Heitman BL, Lessner L (1995) Dietary underreporting by obese individuals - is it specific or non-specific. Br Med J 331: 986-989
16. Martin LJ, Su W, Jones PJ, Lockwood GA, Tritchler DL, Boyd NF (1996) Comparison of energy intakes determined by food records and doubly labeled water in women participating in a dietary intervention trial. Am J Clin Nutr 63: 483-490
17. Sawaya AL, Tucker K, Tsay R, Willett WC, Saltzman E, Dallal GE, Roberts SB (1996) Evaluation of four methods for determining energy intake in young and older women: comparison with doubly labeled water measurements of total energy expenditure. Am J Clin Nutr 63: 491-499
18. Prentice RL (1996) Measurement error and results from analytic epidemiology: Dietary fat and breast cancer. In press. J Natl Cancer Inst

19. Insull W, Henderson MM, Prentice RL, Thompson DJ, Clifford C, Goldman S, Gorbach S, Moskowitz M, Thompson R, Woods M (1990) Results of a feasibility study of a low-fat diet. Arch Intern Med 150: 421-427
20. Henderson MM, Kushi LH, Thompson DJ, Gorbach SL, Clifford LK, Insull W, Moskowitz M, Thompson RS (1990) Feasibility of a randomised trial of a low-fat diet for the prevention of breast cancer: Dietary compliance in the Women's Health Trial Vanguard Study. Prev Med 19: 115-133
21. Willett WC, Sampson L, Stampfer MJ, Rosner B, Bain C, Witschi J, Hennekens CH, Spiezer F (1985) Reproducibility and validity of a semiquantitative food frequency questionnaire. Am J Epid 122: 51-65
22. Block G, Hartman AM, Dresser CM, Carroll MD, Gannon J, Gardner LA (1986) Data-based approach to diet questionnaire design and testing. Am J Epidemiol 124: 453-496
23. Rossouw JE, Finnegan CP, Harlan WR, Pinn VW, Clifford C, McGowan JA (1995) The evaluation of the Women's Health Initiative: Perspectives from the NIH. J Am Med Women's Assoc 50: 50-55
24. Prentice RL, Sheppard L (1995) Aggregate data studies of disease risk factors. Biometrika 82: 113-125
25. Sheppard L, Prentice RL (1995) On the reliability and precision of within and between population estimates of relative risk parameters. Biometrics 51: 853-863
26. Sheppard L, Prentice RL, Rossing MA (1996) Design considerations for estimation of exposure effects on disease risk, using aggregate data studies. To appear, Statist in Med

Commentary on "Bias in observational studies"

Manning Feinleib, Washington, D.C. / USA

This is actually the third conference on the topic of weak associations or, as we may now be calling them, small effects. E. Wynder organised one almost fifteen years ago and a second some ten years ago. The topic I had to discuss ten years ago had almost the same title as today's talk, "Biases and Weak Associations". So I went back to my previous publication and saw that I could say just about the same things now. But I will give that as a reference (Feinleib 1987; Wynder et al. 1982; Sackett 1979) and go on to some other points at the present time.

You are all familiar with the biases listed below. These are the biases that were mentioned in the article in *Science* recently. They are the basis of the criticisms levelled against epidemiologists when their studies produced conflicting results. All I can say is *mea culpa*. I think that we have all encountered situations that have been the potential for these kinds of biases. We realise that these are issues we must take care of when doing observational studies. We teach our students that avoiding these biases is the real challenge of epidemiology and we hope they will design studies in which these biases are minimised.

Some common biases in studying small effects

- Selection
- Selective participation
- Non-participation
- Appropriate controls
- Measurement
- Exposure Recall
- Measurement
- Exposure
- Recall
- Interviewer
- Confounding
- Publication

Listed below are the main areas in which errors can occur:

- Disease specificity
- Etiologic heterogeneity
- Heterogeneous susceptibility
- Relevant exposures

- Effect modifiers

When discussing biases in the context of small effects, we must be aware that we are dealing with complex issues. Even when the major biases are avoided, we must be aware that there are many other potential biases and limitations in the state of art that can trip us up and produce results that do not agree with those obtained by others.

Why do we study small effects? Why do we not always have large effects to study? The problem is that we are dealing with a multidimensional situation, not just the unidimensional one that is sometimes referred to.

One aspect is the problem of disease-specificity. We have come a long way in the last forty years in studying each specific site of cancer individually, rather than studying all cancers as a group. Another example is the nutrition area where we started to make progress when we realised we were dealing with many specific entities that had been lumped together into a global cluster, but which are now recognised to have distinct clinical manifestations and different etiologies. Thus, a risk factor which has a strong association with a specific cancer site or a specific nutritional disorder, may show a weak association in relation to a broader group of cancers or nutritional disorders.

This brings us to the second point: etiologic heterogeneity. The risk factor being studied may be only one of the causes of a disease in the population. If the control group (the "unexposed" group) has a high rate of disease, then other factors are influencing disease occurrence and the factor one is studying may have a relatively small additional effect.

A third reason for weak effects is that there may be heterogeneous susceptibility and resistance in the population. Epidemiologists, other than those in infectious diseases, have not concentrated on this very much. As we get into molecular biology and genetic studies, we will find that the population is not uniform by any means, and that different people exposed to the same factor can have different results due to different innate or acquired susceptibilities. A classic example of this is phenylalanine, where most people can have all the phenylalanine they want, and nothing will happen to them. However, there are a few people with deficiencies in their genes, which will have a tragic effect if they are exposed to diets with phenylalanine.

Another important issue for epidemiologic studies is that of relevant exposure. I think everyone here knows about dose-response curves. However, it seems to me that when we talk about something like cholesterol level, we are talking about a measurement made once or perhaps a few times, maybe a decade or more before the relevant disease emerges, and this is supposed to characterise all the possible relevant exposures to the disease, and be able to predict why one gets a clot at the bifurcation of a coronary vessel on such and such a date many years later. We know that this is a grossly oversimplified approximation of the relevant exposure. Similarly, if you study traffic accidents and ask people "how many miles do you drive a year?", you will probably find a correlation. But you know that this has a very weak association to the specific circumstances that led to a specific traffic accident. I think this is true for most of the exposures that we have been studying: they are very weak, very loose approximations, and we have to find a more specific approach.

We have not said very much about effect modifiers. Confounding is a separate issue which we will get to in the discussion on biases. Effect modifiers are tremendously important. We always have to take into consideration that the exposures we are looking at do not operate in isolation. A lot of other things are going on, and we have to be able to study those also. One of the major advances in elucidating risk factors for coronary artery disease, as done in the Framingham study, was the development of a model - the multivariate logistic model - which allowed for the effects of several risk factors simultaneously. In this model it is evident that the absolute effect of any single risk factor depends on the levels of the other risk factors present.

This brings us back to the recent *Science* article. One thing about the *Science* article is that it does not contain any fact that we did not know before. Its tone is very aggressive; it is a very anxiety-provoking assessment. But the basic fact of how difficult it is to study any disease in a human population, a human community, is recognised. As we develop better instruments to do this, we will do a better job. As H. Armenian pointed out, epidemiology is a young science. They were arguing about whether there was a distinct field of chronic disease epidemiology within our lifetimes. Now we know that not only is there such a field but it is very complex. The genome project gives us one road map of the immensity of the problem we are going to have to face. Every one of those genes, and there are many thousands of them, is going to have its own role in predicting risks for specific diseases and how it interacts with other factors. So in talking about the small effects, we have to realise that these effects may be currently small because they are operating in a very complex environment. If we look at them carefully, we will often find that the more specific our question, the more homogeneous the disease, the more isolated the causal factor which leads to that disease. The more we know about individual susceptibility and the role of other factors, the more likely we will find that we have large and not small effects.

Some concerns we should think about before we embark on a study - I call them pre-study biases - are listed below. We are familiar with the technological aspects of designing studies to avoid biases in the execution of studies. But how many of us have given careful thought to the various choices that come up when we embark on a study? How do we choose the kind of question that is congenial to productive research? We are obviously concerned about what is relevant at our own particular time in history and in the populations that we have available. What is germane today might be totally irrelevant in 20 years. I recall, not too long ago when I was working on the Framingham Study, that we produced a paper on cholesterol levels and heart disease incidence over a 20-year follow-up period. It was rejected because the reviewers felt that at that stage in scientific development, they were no longer interested in total cholesterol levels - they wanted to know the effects of specific lipid fractions. This was a proper concern related to what I called "relevant exposures" above. We answered this criticism by pointing out that Framingham was designed to be a long term study and there were no data in the literature on 20 years of follow-up for total cholesterol. More than twenty years ago, when these measurements were made, one did not know about lipid fractions. The paper was finally accepted.

Pre-study biases

- Choice of questions
- Choice of populations
- Choice of risk variables
- Choice of co-variates
- Choice of outcome variables
- Choice of study design
- Funding sources
- Possible conflicts of interest

The other items on the list may also reflect subtle biases on the part of the investigators and limit the quality of the study. The last two items, funding sources and possible conflicts of interest, have received special attention recently. These are items critics look at when they try to find hidden motives behind the results of a controversial study. As one colleague said very aptly when he was giving a talk, "I will first give you the biases upon which I based my results. " He meant that only half facetiously because he was well aware that when we make choices about the way we do a study, they depend very much upon various personal biases that we have and may not recognise explicitly. But the question that several editors have put forward is whether investigators should reveal their funding sources so that readers can somehow assess whether there is an inherent potential for bias in the results. I personally am ambivalent about this practice. I think that good epidemiologic studies can be done regardless of who paid for the study. We have to do a better job of reassuring our audience that these potential conflicts of interest do not dominate the science we do or the results we report.

Now what recommendations can we make for minimising bias? There is probably a long list of things that we could go over. I would just like to mention a few of them, most of them in the design phase as listed below. As E. Wynder said, it is good to have a medical background. But depending upon the area of research, it would be good to have a PhD in biochemistry, another PhD in molecular science, another PhD in statistics, etc. When you start doing your investigations at the age of fifty, you may be fully qualified but you cannot always rely on that. So as W. Holland suggested, we have to collaborate with colleagues with other subject matter expertise when doing our studies. As R. Prentice can probably verify, drawing samples is a complex area with its own sophisticated techniques and methodologies. Experts in sampling teach us to avoid grab samples and convenience samples because they are likely to be biased. There are established ways of drawing random samples from rosters of the people in the community that will be representative and unbiased.

Selected strategies for minimising bias

- Design
- Subject matter expertise
- Sampling expertise
- Expertise in measurement procedures
- Implementation
- Quality control
- Analysis
- Interpretation and dissemination

Several colleagues have emphasised that improvements are needed in measurement procedures. That is probably the area that has lagged most in developing the armamentarium of epidemiology. We have not devoted enough resources to developing good measurements in general and especially in the cognitive areas. The question of recall bias is paramount in many case-control studies. In only a few areas have we developed methods to validate self-reported information. One of these is in relation to smoking habits. We can now measure cotinine in urine samples. Several studies have shown that if the subject is aware that cotinine levels will be measured, they tend to give more reliable histories of their smoking habits. It might be a strategy to tell people that for a random sample of subjects we will be doing an objective validation. Even if such validation is done on a relatively small sample, it may improve the accuracy of response in general. There are still a lot of things that we have to explore.

One aspect I was concerned about when I was at the National Center for Health Statistics was the cognitive psychological research unit. How to orient respondents about the topics that you wanted to ask about, how to refresh their recall, how to ensure that they have the appropriate timeframe, and, in general, how to try to get them into a mood in which they would be willing and able to give you accurate information. Work is continuing in this area.

There are many other topics we can go through, but I just want to mention quality control. This involves training the interviewers and the laboratory technicians on the critical need of adhering to study protocols, including respecting the integrity of blinded specimens. We must allow adequate preparation to make sure that things are standardised right from the beginning and then maintain an ongoing program to ensure that staffs adhere to the study protocol. Introducing repeat measurements, observing interviewers in the field, inserting blinded specimens and calibrating all instruments regularly are standard scientific techniques that can improve the quality of epidemiologic studies.

I wanted to mention one new idea that I encountered just recently, that is the technique called instrumental variables. As I understand it, this is a technique that one can use to create a pseudo randomised experiment. One tries to use some characteristic of the subjects that is not directly related to the exposure that one is concerned with but is correlated with it and is also correlated with the outcome. Such an instrumental variable may serve to eliminate some potential biases. One clever example that has been reported of such a variable is a study that comes from the experience with the Vietnam war. The method used to draft young American men during the Vietnam war was to select a number of birth dates at random and the inductees were to be called from those born on these dates. A group at the University of California examined the mortality experience in the state of California of men in their thirties - this was about 15 years after the Vietnam war - to see the proportion of them who were born on days that were subject to the draft. Here the instrumental variable was date of birth. The investigators did not actually ask whether or not these men served in the Vietnam war. The important aspect of this technique is that all men born on the selected birth dates were "at risk". There was no extraneous bias due to confounding factors that may have determined who actually served in Vietnam. The investigators found certain differences consistent with long term effects of the war. There are other examples that are starting to emerge in the literature and I suggest that this is one technique that may help eliminate some of the selection biases so dangerous in the usual observational epidemiological study.

References

1. Feinleib M (1987) Biases and weak associations. Prev Med 16: 150-164
2. Wynder EL, Schlesselman J, Wald N, Lilienfeld A, Stolley P, Higgins ITT, Radford E (1982) Conference report: Weak associations in epidemiology and their interpretation. Prev Med 11: 464-476
3. Sackett DL (1979) Bias in analytic research. J Chron Dis 32: 51-68

Small effects: Subgroup analysis and interaction

Anders Ahlbom, Stockholm / Sweden

Introduction

It is common in epidemiologic studies to investigate relationships between exposure and disease, both in the entire study population and in strata or subgroups. Also in situations where the effect in the entire study population, sometimes referred to as the main effect, is absent or small, is it common practice to continue with subgroup analysis. One rationale for looking at effects in subgroups is that there may be segments of the population in which people have an increased susceptibility and in which an effect of the exposure may be strong or at least easier to detect. There may also be methodological differences across segments of the population that make it easier to detect relations in some parts of the population than in others.

Stratification, that is division into subgroups, is routinely done by basic demographic and socio-economic factors such as age, sex, and marital status. Also, other known or suspected risk factors for the disease are commonly used for stratification whenever data is at hand. This stratification is usually done primarily to inform the choice of statistical model, but may at the same time provide knowledge on subgroup effects. Occasionally, stratification is for variables on which information was collected with the specific purpose of testing a hypothesis regarding interaction, that is a subgroup effect.

Although investigating stratum specific results may appear to be a reasonable approach, subgroup analysis are often criticised, or at least the results are interpreted with great caution. The argument for being sceptical towards subgroup analyses is that chance may have played a role in this. The background is that subgroup analysis are sometimes carried out without specified hypotheses, which, together with the often small numbers in subgroups, make it difficult to distinguish subgroup effects from chance associations. Subgroup analysis may therefore lead to the reporting of false positive subgroup effects.

It may be useful to begin this discussion by briefly reviewing some examples of subgroup analysis. Recently, Li et al. (1995) reported on the possibility that electric blankets used during pregnancy were related to urinary tract anomalies in the offspring. No effects in the study population as a whole were found but some effects were observed in a subgroup made up of women with a history of subfertility (Feychting & Ahlbom 1993). This publication is an example in case. The authors interpreted their findings carefully but felt that the results should be reported such that the findings could be tested by others in subsequent studies. The publication was accompanied by an editorial that was somewhat more sceptical towards the subgroup finding, but in essence conveyed the same message (Hatch 1995). Another example is a study on aryl nitrite exposed workers and bladder cancer in which an association is present only in the subgroup of those with the slow acetylator phenotype (Hallqvist et al. 1996). It seems as if this finding has escaped the criticism of being based on a subgroup analysis, perhaps because it was readily interpretable in the context of genetic susceptibility and therefore highly plausible. A third example is a study on residential magnetic field exposure and childhood leukaemia in which it turned out to be a substantially stronger effect in children who lived in one-family homes as compared to children living in apartment buildings (Li et al. 1995). The reason for the stronger effect in the one-family homes is not known, but it is known that the magnetic field measurements are more valid in the one-family homes. Thus, a possible explanation is a higher amount of non-differential magnetic field exposure misclassification in the apartment buildings, and as a consequence an attenuation of the relative risk.

The purpose of this presentation is to provide a structure for dealing with subgroup analysis in situations of weak main effects.

Interaction

The main rationale behind subgroup analysis is that there may be segments of the population with increased susceptibility because the exposure under study may interact with some other environmental or genetic factor. Interaction between two causes of a disease implies that they are both part of one causal pathway that lead to the disease. This is elegantly described by the *pie model* in which each of several pies represents a way of producing the disease and each slice of the pie represents a risk factor for the disease (Rothman 1976). Slices from the same pie represent risk factors between which there is interaction: adding one slice carries a risk if the other slices in the pie are present but not otherwise.

Small Effects

It follows immediately from the definition of interaction and an examination of the pie model that, whether a risk factor has a small or a strong effect depends on the prevalence of the interacting risk factors, that is of the prevalence of other risk factors in the same pie. As an example, consider an environmental risk factor that interacts with a gene, such that people with the gene are susceptible to the risk factor and therefore likely to develop the disease when exposed. If in one population this gene is rare, the effect of the risk factor will as a consequence be small, while it may be strong in a population in which the gene is common. Similarly, an infectious agent may have an effect only on people with a weak immune defence system and therefore the effect of exposure to the infectious agent would depend on the proportion in the studied population with the weak immune defence system.

Although it does not follow immediately, it is easy to show that interaction as defined in the pie model is manifested, in terms of disease rates, as departure from additivity (Shields & Harris 1991). That is, with no interaction between two risk factors, the effect of combined exposure is the sum of the two separate effects; with (positive) interaction the effect of a combined exposure exceeding additivity. A multiplicative effect for those with joint exposure to two risk factors would therefore usually indicate interaction between the risk factors. This may have implications for the interpretation of subgroup analysis when the magnitude of the effect is measured as the relative risk. If the base-line rate of the disease varies across subgroups, the relative risk will be small in subgroups with high base-line rates and big in subgroups with low background rates even with no interaction at hand. Thus, the possibility of varying base-line rates must be factored, whenever relative risks are used.

Subgroup Analysis

Since a small effect of a risk factor in a study population may be due to a low prevalence of a factor that interacts with the risk factor, it is reasonable to assume that there may be a subgroup of the study population in which the effect of the risk factor is strong because of a higher prevalence of the interacting factor in this segment of the population. The identification of such a subgroup is useful when assessing whether or not an exposure is indeed a risk factor for the disease. Depending on the type of variable that was used to define the subgroup, its identification may also shed light on the mechanisms through which the risk factor asserts its effect. As pointed out already, not all high subgroup effects are, however, due to interesting interaction phenomena. There is also chance to consider. When the relative risk is used, the possibility of varying base-line incidence of the disease must also be considered.

Perhaps the most intriguing issue is the distinction of interaction, that is identification of a susceptible subgroup, from chance variations. The introductory example of electric blanket use and birth defects is a rather typical example in that both the authors of the publication and the author of the editorial were uncertain about the interpretation. In particular, they warned that the high effect in the subgroup with a history of infertility could be a chance finding. This possibility has also been raised in relation to the residential magnetic field exposure example with a higher effect in one-family homes. Most readers, however, seem willing to accept the explanation of lower degree of non-differential magnetic field exposure misclassification in one-family homes. In relation to the aryl nitrite example with the high effect in the slow acetylator, subgroup interaction and a susceptible subgroup is readily accepted as an explanation.

90

The different ways these examples have been interpreted may in fact provide some guidance on how to deal with subgroup results. One difference between the three examples is the amount of credibility given to the proposed explanation. The principles for interpretation of subgroup results are in general the same as those that guide the interpretation of main effects. In brief, the interpretation of any epidemiologic finding, main effect or subgroup effect, is based on the pooling of 1) the statistical stability in the effect estimate; 2) the amount of support from previous epidemiologic studies, and 3) the biomedical plausibility assessed from experimental research and other sources of information. There are two circumstances that set subgroup analysis somewhat apart. First, the number of subjects in the various subgroups is often small, which leads to unstable estimates and consequently to an increased likelihood of risk elevations due to chance. Second, subgroup analyses are commonly done for reasons other than the identification of susceptible subgroups. For example, when controlling for confounding, homogeneity has to be checked across strata in order to inform about appropriate statistical models. In this process, strong effects in specific strata may be detected without there being a biomedically plausible explanation at hand. Yet, the same principles apply as in main effect analysis, but often the small numbers and the also common lack of known biological explanation make it particularly important to consider the possibility of a false positive results.

Even though these main principles are quite clear, a strategy on how to handle small group effects in practice remains to be developed. Should one only report subgroup results when a hypothesis regarding interaction was specified a priori? Is it sufficient that one can come up with a reasonable explanation when the subgroup result has been observed? Should a subgroup finding be supported if it is sufficiently statistically stable even when there is no plausible explanation at hand?

If the only concern was to avoid false positive reporting of subgroup findings, the situation would be rather simple. One need only to apply sufficiently strict criteria for what to report and the risk of false positive findings would be minimised. The problem, however, is that some subgroup findings may carry significant information that one does not want to miss and that may turn out to be important, if confirmed by subsequent investigations. In other words, there is also a concern regarding false negative results. The challenge is to find the proper balance.

References

1. Feychting M & Ahlbom A (1993) Magnetic fields and cancer in children residing near Swedish high voltage power lines. American Journal of Epidemiology 138: 467-48
2. Hatch M (1995) What can we infer from findings in subgroups? Epidemiology 6: 473-475

3. Hallqvist J, Ahlbom A, Diderichsen F, Reuterwall C (1996) How to evaluate interaction between causes: a review of practices in cardiovascular epidemiology. Journal of Internal Medicine 239: 377-382
4. Li DK, Checkoway H, Mueller BA (1995) Electric blanket use during pregnancy in relation to the risk of congential urinary tract anomalies among women with a history of subfertility. Epidemiology 6: 485-489
5. Rothman KJ (1976) Causes. American Journal of Epidemiology 104: 587-592
6. Shields PG, Harris CC (1991) Molecular epidemiology and the genetics of environmental cancer. Journal of the American Medical Association 266: 681-687

Enhancing Small Risks in Epidemiologic Studies

Lenore Kohlmeier, Chapel Hill /USA

Small risks in epidemiology are defined by some as relative risks below 1.2, 1.5 or even 2. However, even odds ratios of 1.001 can be significant. Since an odds ratio of 1.001 can be significant, when continuous measures of exposure are used in analyses-arbitrary cut offs for "relevant" relative risks are meaningless. When epidemiologists stop using quintiles and modeling with continuous variables, the range of measurement will determine both risk and relevance. A relative risk is by definition the risk of a given exposure at a defined time period in a given population relative to a defined disease at an arbitrarily defined baseline. This being the case "small risks" can be enhanced by lowering the baseline.

The question is not really one of small risks, it is a question of real risks. Are we measuring noise or are we measuring something real? And perhaps equally important, are these relevant risks? Are we spending our time where we should be as public health professionals? This paper will not address the latter, but rather the possibilities to enhance small risks through epidemiologic study designs.

At least three strategies exist to increase small risks: Lower your baseline, enhance the variance and exposures (selection of exposure extremes to improve your measurements), and use susceptibility markers.

Lowering the baseline

Generally, control populations are selected to randomly represent the population at large, leaving the distribution to chance. This distribution is divided up in quintiles for the purpose of adequate cell sizes and examined for trends. A more proactive approach to determining the shape of the distribution, and number of cases in the exclusions is being used in some trials. Figure 1 shows this distribution. And if one could move the extreme quintiles further apart, a greater gap between the baseline and top quintiles would arise. One strategy to achieve this is to design studies based on more rectangular distributions. In comparison to the normal distribution cutoffs, more people would be in the extremes, and the distance between cutoffs would be greater (Figure 2). An example of this is that being used by the National Cancer Institute in the US in the American Association for Retired Persons' Study. To look at dietary fat and its effect on morbidity from a number of diseases, two and a half million people were screened and only those from the extremes of fat intake maintained as a basis for the cohort. This is to maximise power to compare extremes. Another approach to this is what we have done in the

EURAMIC Study which compares women across Europe with a greater variety of diet rather than a population from a more homogenous region or a single occupational group. This provides, in a similar fashion, a broader distribution and more distance between extreme quintiles.

Figure 1

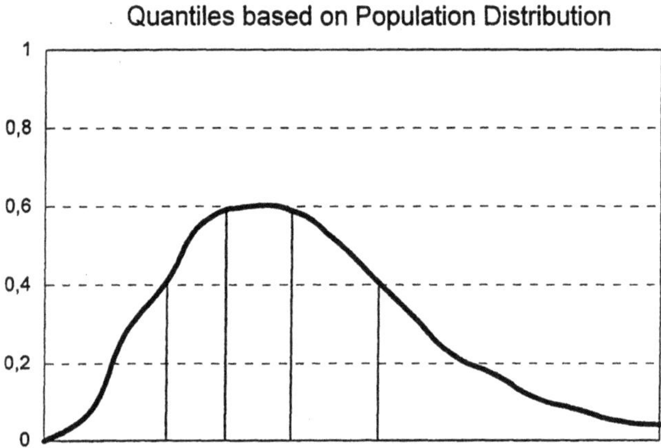

Figure 2

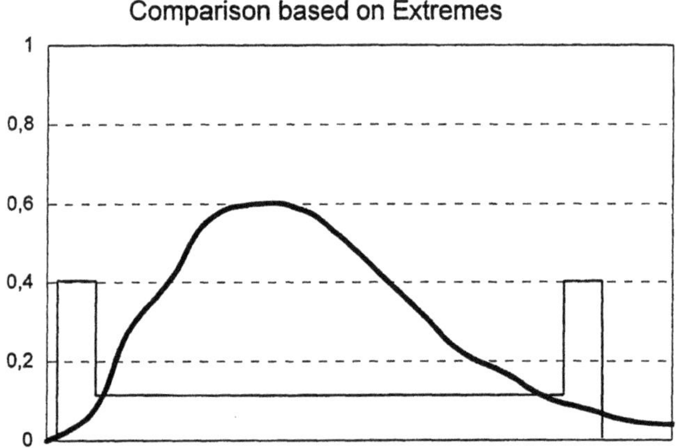

Use of continuous variables

Usually quintiles and their cutoffs in cohorts are used to define the comparison groups. In fact, most analysis are done to determine the change in risk with a given change in exposure level. Only then can results be translated into a public health message. When exposures are modeled as continuous variables, the odds ratio will be per unit increase as measured. In our studies on adipose tissue levels of micronutrients this may translate into the increase in risk per microgram of vitamin E stored in each gram of fat. In this case 25 and 75% contrasts can be used to get a sense of the magnitude of benefit in moving from one end of the exposure distribution to the other. Use of 10th and 90th percentile cutoffs also serve to enhance risks by increasing the distance of the comparison.

Susceptibility markers

Susceptibility markers include but are not limited to genetic markers. Other susceptibility markers can change, such as age or whether or not you've been exposed to a virus. However, currently, susceptibility markers in epidemiology refer mainly to polymorphisms which are prevalent in the population and can affect risk. In the area of carcinogenesis, polymorphisms might express a turned-on oncogene, a turned-off tumor-suppressor gene or perhaps a polymorphism of a carcinogen-activating enzyme or a polymorphism of a detoxifying enzyme. Thanks to the advent of PCR, we are increasingly able to measure these in small samples of DNA. After the not uncomplicated issues of informed consent for conducting such analysis are resolved, this opens opportunities for the detection of high risk groups. Fortunately for the epidemiologist, many of the polymorphisms of metabolic enzymes currently under study show high penetration levels in the population, which allows adequate or optional cell sizes in all groups in large studies.

Availability of information on polymorphisms affect the traditional epidemiological models of exposure and disease by allowing subgroup analysis. By sorting individuals into those that are polymorphism-negative, one might detect hidden risks. This potential calls for new, more efficient and more powerful study designs in epidemiology which allow exploration of these types of interactions between genetic polymorphisms and exposures.

One approach is to design studies in which only the more susceptible groups are included. An example of this can be seen in the relationship between NAT genotype and colon cancer risk (Bell et al. 1995b). The data in table 1 show the population penetrance levels, and the risk of colon cancer for individuals within each of these groups. In this North Carolinian population of mixed races 45% are NAT2 rapid acetylators, 55% slow acetylators. If you are a 1*11, you have a lower risk of colon cancer than if you are 1*4. If you are 1*10, the risk is 80% greater. NAT2 rapid acetylator status does not seem to carry an independent risk, but if you have both NAT2 as a rapid acetylator and the NAT1*10, then the odds ratio is 2.8. This differs somewhat from population to population.

Table 1
Odds Ratios for Colon Cancer by NAT Status

Nat1*11	0.6	6%
Nat1*4	1.0	64%
Nat1*10 homozygote	1.8	3%
Nat1*10 heterozygote	1.8	27%
Nat2 rapid	1.1	45%
Nat1*10 & Nat2	2.8	

Bell et al., Cancer Research 55:3537 1995

Case-only studies

The study of interactions between genes and an exposure can be conducted through case-only studies in which only cases are evaluated for genotype and prior exposures. Germ-line DNA is required for determining genotype. This is an extraordinary marker as it is not vulnerable to the recall or information bias which often plagues case-control studies. Case-only studies have been proven independently by two groups of statisticians to be efficient designs for the study of interactions (Piegorsch et al. 1994. Begg and Zhang 1994). They are based on the principle that the odds ratio for some disease depending on your genetic phenotype and your exposure levels is a function of the risk in the cases versus the risk in the controls, which will turn out to be equal to the risk in the cases only. Begg and Zhang have published on the use of case-series methodology to examine genetic makers of disease progression in a DNA marker that occurs with the progression of tumors, which is basically the same method in a different context. They consider how the exposure may influence a specific mutation, rather than how an existing allele may influence sensitivity to disease, and disease related exposures.

There are a few caveats in using case-only methodology for studying interactions. One is an assumption of independence of exposure and genetic susceptibility. For example, in studying a mixed gender, or a mixed race population, groups may differ in prevalence of genetic polymorphism exposure. For example, as well as differing in NAT1 genotype incidences, African-Americans may also eat more bacon and grilled meat. In this case, the assumption of independence of the genetic marker and the exposure is violated. The second assumption is a rare disease assumption: less than 20% of the population should be afflicted.

The benefits of this study design or analytic approach involves increased precision for detecting interactions, lower standard errors, and a much lower cost of the studies due to the fact that cases only are needed. Our calculations show that the number of cases may be 40% lower than needed in a full study of main effects (Kohlmeier et al., 1997).

In conclusion: 1) the contrast defines the magnitude of risk; 2) small risks appear larger if the baseline for comparison is lower or there are more individuals in the extremes of the distribution; and 3) susceptibility markers, when used appropriately, may contribute to strengthening risk. An appropriate use of this information is in the context of the case-only method, which offers tremendous precision for determining whether or not an interaction exists.

These risk enhancing methods are appropriate only in the context of determining whether a risk truly exists. With improved measurements and studies richer in information, epidemiology will become a more quantitative science, providing us with the more desired outcome: what the increase in risk is per unit increase in exposure. From there the relevance in the population can be derived, and the costs and benefit basis for decisions on altering risk levels soundly developed.

References

1. Begg CB, Zhang ZF (1994) Statistical analysis of molecular epidemiology studies employing case-series. Cancer Epidemiology, Biomarkers & Prevention 3: 173-175
2. Bell DA, Badawi AF, Lang NP, Ilett KF, Kadlubar FF, Hirvonen A (1995a) Polymorphism in the N-acetyltransferase 1 (NAT1) polyadenylation signal: Association of NAT1*10 allele with higher N-acetylation activity in bladder and colon tissue. Cancer Res 55: 5226-5229
3. Bell DA, Stephens EA, Castranio T, Umbach DM, Watson M, Deakin M, Elder J, Hendrickse C, Duncan H, Strange RC (1995b) Polyadenylation polymorphism in the acetyltransferase 1 gene (NAT1) increases risk of colorectal cancer. Cancer Res 55: 3537-3542
4. Clayton EW, Steinberg KK, Khoury MJ, Thomson E, Andrews L, Kahn MJE, Kopelman LM,Weiss JO (1995). Informed consent for genetic research on stored tissue samples. JAMA 274:1786-1792
5. Kohlmeier L, DeMarini, Piegorsch WW (1997) Nutrient Interactions In Nutritional Epidemiology. In: Design Concepts in Nutritional Epidemiology. Oxford Press
6. Piegorsch WW, Weinberg CR, Taylor JA (1994) Non-hierarchical logistic models and case-only designs for assessing susceptibility in population-based case-control studies. Statistics in Medicine 13: 153-162
7. Trials 13: 170-177
8. Ad Hoc Working Party of the International Collaborative Group on Clinical Trial Registries (1993) Position paper and consensus recommendations on clinical trial registries. Clin Trials Meta-Anal 28: 255-266
9. Bero L, Rennie D (1995) The Cochrane Collaboration. Preparing, maintaining and disseminating systematic reviews of the effects of health care. JAMA 274: 1935-1938

Is meta-analysis a valid approach to the evaluation of small effects?

Samuel Shapiro, Boston / USA

In this presentation, meta-analysis is defined as the synthesis of data across studies in order to produce combined estimates of effect. (The term has also been applied to the quantitative description of how an effect and its determinants may vary among studies. That aspect of meta-analysis is not considered here). Meta-analysis was introduced into medical research about a decade ago. There has been an explosive growth in its use, particularly in the evaluation of small effects. Initially, meta-analysis was confined to the synthesis of randomised controlled trials (that application is not considered in this presentation), but methodology was soon extended to the evaluation of nonexperimental (observational) data.

A further and important subdivision of the methodology has been between the meta-analysis of published data only, as against the meta-analysis of the "raw" data from the included studies (an approach that has also been described as "combined analysis", "pooled analysis" or "overview"), with or without the active collaboration of the original investigators. The procedure often involved the redefinition of variables, "data trimming", respecification of class intervals, and other procedures designed to render the data sets compatible and combinable.

In tandem with the growth of meta-analysis, statistical methods have been developed to assess the heterogeneity of any given effect among studies, and to develop appropriate methods of data aggregation. In this presentation the statistical issues will not be considered: they are assumed to be valid. What will be considered is the validity of meta-analysis from the epidemiological perspective.

This conference is concerned with the problems of selection bias, information bias, confounding, misclassification, timing, subgroup effects, consistency and generalisability that bedevil the evaluation of small effects. Against that background the validity of meta-analysis as applied to nonexperimental (observational) data is considered. The conclusion reached in this review is that meta-analysis in that domain is not valid.

The Robert Koch Institute is a uniquely appropriate setting for the consideration of causality as applied to small risks, and for that reason I am happy to be here. Koch was the first person to develop a set of causal criteria. His purpose was to apply them to infections, and it was his pioneering efforts in that application that subsequently stimulated Hill, Lilienfeld and Susser, among others (1991), to propose corresponding criteria for application to epidemiologic research.

One important criterion of causality is the strength of any given association. In any reasonably well conducted study, a weak association may be due to confounding or bias, but it is unlikely that a strong association can be completely explained away by defects in study design. That point is critical to the topic of meta-analysis: when associations are strong (as, say, with smoking and lung cancer), there is no need to resort to it. It is when associations are weak that meta-analysis are tempted to combine studies, in the erroneous belief that the statistical significance thereby accomplished translates to causality.

At E. Wynder's request, some of the material that will be considered today has been published before, but I will refer again very briefly to it in order to systematically cover the general theme. I would like to commence by referring to a disastrous error in which I was a participant at a relatively young epidemiological age (Boston Collaborative Drug Surveillance Program 1974). That error has influenced my attitude to the process of inferring causality from nonexperimental data ever since. In 1974 the first study in Figure 1 stimulated the hypothesis that rauwolfia alkkaloids increase the risk of breast cancer (Boston Collaborative Drug Surveillance Program).It was published back to back with two other studies (Armstrong et al. 1974, Heinonen et al. 1974) in a single issue in *Lancet*. All three studies were acknowledged at the time to have methodological limitations, but based on the total evidence, the respective authors felt that the defects in the individual studies, as it were, "cancelled each other out". The three studies, in their turn, stimulated a cascade of additional studies (World Health Organisation 1980).

In 1979 the International Agency for Research on Cancer reviewed 15 studies (WHO 1980) (Figure 1), and concluded that the evidence from the better conducted ones was against an increase in the risk of breast cancer attributable to rauwolfia. Finally, yet another study (Shapiro et al. 1984) produced statistically stable data that showed no increase in the risk of breast cancer associated with rauwolfia use: the relative risk was 0.9, and the study was large enough to set an upper 95% confidence limit of 1.2.

The hypothesis-raising study (Boston Collaborative Drug Surveillance Program 1974) was based on 150 cases of breast cancer, and the final, null study (Shapiro et al. 1984) was based on 1816 cases. In their essentials the methods were the same in both studies (Boston Collaborative Drug Surveillance Program 1974, Shapiro et al. 1984). There were never adequate grounds to propose the hypothesis in the first place, and the original association was probably due to chance, even though it was statistically significant. Yet, as one scans the data in Figure 1, it is a reasonable bet that has a meta-analysis been undertaken at the time, it would have produced a positive overall association that could well have been labelled as "causal". Fortunately, when these studies were published the technique of meta-analysis had not yet been applied to medical data.

Figure 1
Risk of breast cancer in relation to reserpine use: relative risk estimates and 95% confidence intervals in 13 case-control and two cohort studies. Confidence intervals are not provided if they were not published or could not be estimated from the published data. Some studies estimated more than one relative risk.

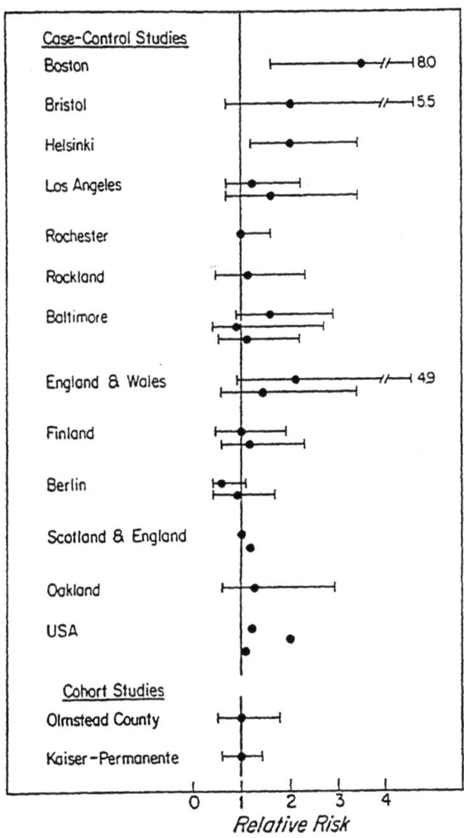

[Note: This figure has previously been published in reference 10.]

These thoughts occurred to me when the late T. Chalmers and his group published one of the early meta-analysis of randomised controlled trials (RCTs). They suggested that anticoagulant therapy improves the prognosis of myocardial infarction (Chalmers et al. 1977) - a conclusion that could not confidently have been reached simply from an assessment of the data in any individual study. I suggested to Chalmers that the new methodology might be of some use in combining information from RCTs, but that it could not be used as a valid tool in non-experimental research because of uncontrollable confounding and bias. He disagreed, and soon moved on to the meta-analysis of nonexperimental data as well (Longnecker et al. 1988).

Meta-analysis rapidly achieved considerable popularity - so much so that it is now unusual to open a medical journal that does not contain one. Today we can further broadly divide meta-analysis into four types: the meta-analysis of published nonexperimental data, of "raw" nonexperimental data (sometimes the preferred terminology is "combined analysis", "pooled analysis", "collaborative analysis" or "overview"), of published RCTs, and of "raw" data from RCTs.

In the critique that follows, I will not consider the meta-analysis of RCTs. RCTs are only exceptionally designed to identify causes of disease. Because of randomisation, confounding is relatively unlikely in any individual RCT and exceedingly unlikely in a meta-analysis. In well-designed RCTs bias can either be eliminated, or else controlled much more rigorously than is possible in non-experimental research. There are nevertheless problems in the meta-analysis of RCTs (for example, variable quality; generalisability), but the subject is large and complex, and it is best dealt with as a separate topic.

One of the first meta-analysis of published non-experimental research was an evaluation of breast cancer in relation to alcohol consumption (Longnecker 1988). From the published data there were statistically significant summary relative risk estimates of 1.4 for the case-control studies, and 1.7 for the cohort studies - both of them "small" risks. Elsewhere, I have reviewed the findings of that meta-analysis and of a subsequent updated one (Longnecker 1994). Here is a summary of the main defects. There was misclassification and variable definition of alcohol consumption among the studies; there was misclassification of the timing of intake and of the quantity consumed. Multiple sources of information and selection bias were likely, and quite possibly in the same direction across the studies. A spurious "quality score" was used to assess the individual studies. Multiple sources of confounding were present in all the studies. Particularly with regard to the latter possibility, the determinants of the changing incidence of breast cancer are still largely unknown. Many of the determinants of alcohol consumption are still largely different in different cultures, in each of which they are also changing and difficult or impossible to measure. Yet, despite such considerations, the authors suggested that the relative risk estimates derived from their meta-analysis were consistent with causality.

When my critique of the alcohol/breast cancer meta-analysis was published (Shapiro 1994), one of the authors acknowledged that mistakes had been made, but he argued that the technique was then still in its infancy, and that the methodology had since improved (Longnecker 1995, Shapiro 1995). That claim is tested in the next example, a meta-analysis of 12 studies of cancer risk in relation to water cholorination (Morris et al. 1992), again carried out by Chalmers' group, several years later. The following were the main results (Table 1): for all the cancer sites studied the relative risk estimate was 1.15, and statistically significant; for bladder cancer it was 1.21; for rectal cancer 1.38. Applying these estimates to United States incidence rates, the authors claimed that at least 4,200 cases of bladder cancer per year and 6,500 cases of rectal cancer per year are associated with water chlorination.

Table 1
Water chlorination and cancer* - meta-analysis of 12 studies

Site	RR	(95% CI)
All	1.15	(1.09-1.20)
Bladder	1.21	(1.09-1.34)
Rectum	1.38	(1.01-1.87)

*see reference 13

In this meta-analysis, an attempt was again made to quantify the quality of the individual studies. Studies were scored on the basis of selection of subjects, measurement of and adjustment for confounding variables, exposure assessment, and statistical analysis. The overall quality score was calculated from the three subscores: a general methods score, a data analysis score, and an exposure assessment score. Each subscore was calculated as the percentage of applicable quality criteria that were met in each study. The cumulative quality score was a weighted average of the three scores, with both general methods and exposure assessment receiving twice the weight of the data analysis score.

Such language, I believe, is sufficiently dense and arbitrary to confirm what we all know to be true: that quality cannot be scored, measured and taken into account. Moreover, who are these meta-analysts sitting on high, to decide for the rest of us what is and is not good quality and then to measure it? Quality is best evaluated qualitatively: as opposed to meta-analysis. In any adequate qualitative review, we require that the author should give reasons for judging the quality of any given study as good or bad in transparent and easily comprehensible language. It is then up to the reader to decide whether he/she agrees or disagrees.

It is nevertheless instructive to examine the results of the quality assessment (Table 2). Among the 12 studies the median score for all of 25 quality criteria specified as desirable by the authors was only 55%, and it was as low, or lower, for most of the subscores. In addition, depending on the criterion, the authors were only able to assess quality for 6 or 7 of the 12 studies. In other words, by the authors' own standards, the data were unsatisfactory - so unsatisfactory that the only sensible thing to do, surely, would have been to abandon the enterprise.

But it was not even necessary to undertake the impossible task of quantifying the quality of the individual studies, because there was clear evidence of bias in the presented data. Those studies that reported elevated risks tended to report elevations for all cancers studied, and not for specific cancers. To illustrate that point, in Table 3, two of the largest studies are compared (Alavanja et al. 1978, Brenniman et al. 1980). Alavanja et al. (1978) studied 7 cancer sites, and found elevated risks associated with chlorine exposure for all of them, ranging from 1.61 to 2.12. By contrast, Brenniman et al. (1980) found relative risks in the range of 0.97 to 1.22, all but one of them (a point estimate of 1.13 for colorectal cancer) compatible with unity. With the possible exception of x-rays, there is no known carcinogen that increases the risk of cancer at all the sites listed in Table 3. Nor is it likely that there can be. The epidemiologic characteristics of the various tumour sites are markedly different, and there is no biologic plausibility for a universal carcinogen that would increase the risk across the board. By that standard, the study of Alavanja et al. (1978), which made a major contribution to the overall findings in the meta-analysis, was clearly biased.

Table 2
Water chlorination and cancer - quality scores for specific criteria*

	Studies complying %	
Criteria	Low	Median
Selection (n=6)	40	75
Adjustment for confounding (n=7)	10	50
Exposure assessment (n=6)	0	50
Analysis (n=6)	10	55
Total (n=25)	0	55

*see reference 13

Based on this example, the claim (Longnecker 1995) that the meta-analysis of published studies has emerged from its infancy, and improved its methods, cannot be defended.

Table 3
Water chlorination and cancer* - relative risks for gastrointestinal and bladder cancer deaths in two studies

	Alvanja*	Brenniman†
Bladder	1.69‡	0.98
Colon	1.61‡	1.11
Colorectal	1.71‡	1.13‡
Esophagus	2.12‡	0.97
Liver	-	1.00
Pancreas	1.97‡	1.02
Rectum	1.93‡	1.22
Lung	1.79‡	-

* see reference 14   †see reference 15   ‡lower 95% confidence interval excludes 1.0

I turn next to the meta-analysis (or combined analysis) of "raw" data. The conceptual argument for such a procedure is that the published information may not be sufficient to conduct a valid meta-analysis; that it is commonly necessary to record and otherwise resort variables across studies to make them compatible, and to make other adjustments before we are able to properly synthesise information. That objective cannot be accomplished by a meta-analysis confined to the published material.

One of the first major exercises along these lines was conducted by Howe et al. (Howe et al. 1990), who initially attempted to examine 14 case-control studies of breast cancer risk in relation to dietary fat intake. Two studies had to be excluded because the authors declined to collaborate (which, incidentally, casts doubt on the validity of the meta-analysis, ab initio). Depending on the specific dietary element under study, three to four of the remaining studies were then excluded because they exhibited markedly heterogeneous effects, so that in the end, only 8 to 9 of the original 14 studies were meta-analysed.

The principal findings were as follows (Table 4): the relative risk estimate for the highest quintile of total fat intake was 1.46, and for saturated fat it was 1.57. The associations were statistically significant. The main conclusion was that from a combined analysis of international data a high intake of fats, saturated fats in particular, appear to increase the risk of breast cancer by some 1.5-fold.

Table 4
Dietary fat and breast cancer*

	No. of studies	Relative risk Q5 vs, Q1†	P trend
Total fat (g/day)	8	1.46	0.0002
Saturated fat (g/day)	9	1.57	<0.0001
Vitamin C (mg/day)	8	0.86	0.031

*see reference 16
†Quintiles

Now, based on ecological and other evidence, it is reasonable to propose that dietary fats may indeed increase breast cancer risk, but the question here is whether the meta-analysis meaningfully tested that hypothesis. The following is a partial list of some of the problems that were present in this study.

Dietary fat intake is notoriously difficult to record in interview-based studies, with correlation coefficients relative to "gold standard" measurements (such as from prospectively recorded food dairies) usually of the order of 0.3 or less, and seldom better than 0.5 (Shapiro, in press). In this instance, the misclassification thereby introduced was further compounded by having to combine heterogeneous questionnaire data. For example, some of the data were based on a 24-hour dietary recall instrument, and others on periods covering as much as two weeks. Some questionnaires covered as few as 22 food items, others as many as 80. Such misclassification could easily have facilitated the occurrence of information bias. The hypothesis was widely broadcasted, and differential reporting of fat intake could have been ubiquitous across the studies. Probably there was also uncontrollable confounding. For example, fat intake may have been similarly associated with the socio-economic status in the various studies.

If such considerations were not sufficiently daunting, the investigators simply dealt with heterogeneity by means of circular reasoning. They made the remarkable claim that the relative risk estimates among the studies were homogeneous (Shapiro in press). Of course they were and had to be, since those studies that exhibited heterogeneity in the first place were excluded.

(Author's note: since the Potsam conference, a further meta-analysis of breast cancer risk in relation to fat intake that included seven follow-up studies has been published with null results (Hunter et al. 1996). Here we are confronted, not for the first time, by the ultimate irony of two conflicting meta-analysis, neither of which enlighten us as to whether saturated fats do or do not influence breast cancer risk.)

There has been yet a further refinement to the idea of meta-analysing "raw" data. The argument has been advanced that the objective can only be achieved if the original investigators are themselves involved in the nuts and bolts of the enterprise. Another example is a combined analysis of 12 case-control studies (Whittemore et al. 1992, Harris et al. 1992) of ovarian cancer where this was done. The meta-analysis (or collaborative analysis, the designation preferred by the authors) was carried out over a four-year period by one group, which met each year for several days with representatives from the original 12 studies, all of whom are experienced epidemiologists. The latter were intimately involved with the manipulations to which their data were subjected, with the analysis, and with the interpretation of the results.

The collaborative analysis excluded four American studies (Annegers et al. 1979, Demopoulos et al. 1979, Newhouse et al. 1977, Wynder et al. 1969) on the grounds that they did not involve personal interview (Annegers et al. 1979, Demopoulos et al. 1979, Newhouse et al. 1977), or that the data were not available in computer-retrievable form (Wynder et al. 1969), which incidentally, are not valid reasons, and three European studies that were meta-analysed and published separately (Negri et al. 1991), which, incidentally, are also not valid.

Some of the findings from the American collaborative analysis documented the obvious. For example, the high risk associated with nulliparity and the protective effect of increasing parity. Those associations were so strong that they were fully demonstrable in each of the individual studies. For that purpose, no meta-analysis was required. A further claim made for the collaborative analysis, however, was that it was possible to evaluate the separate effects of nulliparity, as opposed to infertility, as independent risk factors for ovarian cancer. That claim is questionable, since infertility and nullparity were so closely comingled and correlated that it is questionable whether any distinction that was observed was biologically meaningful. Elucidation of the question of whether infertility, independently of low parity, as a risk factor for ovarian cancer could require a study designed along lines radically different from the case-control studies included in the collaborative analysis.

A further claim was that analysis of the risks according to whether hospital or community controls were used yielded somewhat different results. The amplified claim was that the studies that enrolled community controls were superior to those that enrolled hospital controls. If so, why were the inferior studies included? By contrast, in a qualitative review the author could have considered the validity of the control selection in each individual study.

In short, this most "ideal" of meta-analysis yielded no new information regarding parity, a well-documented and powerful risk factor. For that factor we are not in the domain of small risks, and a meta-analysis was not needed in the first place. Nor was the meta-analysis able to sort out the separate risks related to parity and infertility.

Setting those matters aside, however, a major claim made to justify the collaborative analysis was that it uncovered one new and important association that would not otherwise have been discovered in the individual studies (Table 5). Three of the 12 studies (Hartge et al. 1989, Cramer et al. 1983, Nasca et al. 1984) that recorded information on the receipt of fertility drugs. In the combined data (Whittemore et al. 1992) there were 626 cases of epithelial ovarian cancer and the relative risk was significantly elevated at 2.8. In the subgroup of mulligravidae it was 27 and among gravidae it was 1.4 and non-significant. Some of the cancers (numbers not given in the text) were classified as tumours of low malignant potential (Shapiro 1994). The relative risk for those tumours was 4.0 and statistically significant.

Table 5
Use of fertility drugs in relation to risk of epithelial ovarian cancer*

	Number of cases	Number of exposed	RR (95% CI)
Invasive			
Nulligravid	88	12	27 (2.3-316)
Gravid	538	8	1.4 (0.5-3.6)
Total	626	20	2.8 (1.3-6.1)
Low malignant potential			
Total	?	?	4.0 (1.1-13.9)

* see references 19 and 20

The authors interpreted the associations as supporting the hypothesis that fertility drugs increase the risk of ovarian cancer. Yet, as shown below, it is readily demonstrated that whatever did account for the findings, it was not the use of fertility drugs.

Firstly, as expected, well over 50% of the cases were over 50 years of age when they contracted ovarian cancer in the late 1970s. Infertile women would thus have sought medical help at the age when they needed it, mostly in the early 1960s, or earlier. At that time, fertility drugs either did not yet exist in the United States (e.g. clomiphene), or they had only recently been experimented with on a small scale in preliminary and exploratory studies (e.g. menotropins) (Shapiro 1994). As a matter of simple arithmetic, only an odd case or two that occurred at an unusually young age could have been exposed to fertility drugs.

That arithmetic was confirmed in subsequently published data (Shapiro 1995) from the three studies (Table 6). In the combined series of invasive cancer and tumours of low malignant potential, there was only one case known to have been exposed to a fertility drug. That patient was a 30-year old women who received clomiphere a short time before she was diagnosed as having ovarian cancer. The strong likelihood is that the cancer or a precursor lesion "caused" the infertility, and its treatment, not the reverse. That single case apart, the remaining drugs that were identified were not fertility drugs. Most of the drugs were unknown, in large part because in one study (Nasca et al. 1984) the names of the specific drugs were not recorded. However, the only defensible assumption that can be made about the unknown drugs, based on distribution of the known exposures shown in Table 6, and on the timing of the exposures, is that they were not fertility drugs.

Table 6
Fertility drugs used by cases of ovarian cancer*

Drug	Number
Clomiphene	1
Estrogens	4
Estrogen and progestogen	1
Thyroid hormone	2
Dextroamphetamine and amobarbital	1
Unknown (invasive cancer)	16
Unknown (low malignant potential)	?

* Data provided to the author by the authors of references 26, 27 and 28

Now, there are good experimental and clinical grounds to suggest that fertility drugs do indeed increase the risk of ovarian cancer (Whittemore 1994). The point made here, however, is that the combined analysis made no contribution to answering that question.

In summary, this most sophisticated of combined analyse produced only one new finding, and a demonstrably false one. How could such an obvious and avoidable error have been committed by an experienced group of investigators? Part of the answer must be that even when investigators actively collaborate in a meta-analysis under the most optimal conditions, it is impossible for them to be as immersed in the data as they would be if they were engaged in the analysis of their own studies. The error demonstrated yet another defect of meta-analysis. By definition, such an undertaking is conducted at one or more removes from the original data. The greater the number of removes, the greater is the likelihood of error, due simply to an increasing lack of familiarity with the intricacies of the study material.

My final example has been selected to examine a general theme that pervades the meta-analytic literature. The argument is as follows: in the meta-analysis of a large number of reasonably well-conducted studies, bias and confounding should, in the aggregate, tend to "cancel each other out" - as has been stated or implied in some of the studies reviewed above. That argument tend to be made more explicitly for confounding; and when it is applied to RCTs, it is undoubtedly true. For the argument to hold true in the domain of non-experimental research, however, the very large and dubious assumption must be made that the right studies, with the right weights, in the right decisions, are present. Otherwise, the "cancelling out" will not occur. Even if it is assumed that there is no bias, and that uncontrolled confounding is the only issue, there can be no reassurance that the "cancelling out" will occur, since the same confounder may be shared by more than one study.

That point has been quantitatively illustrated by Postuma et al. (1994) who carried out a meta-analysis of studies of oestrogen use in relation to the occurrence of either cardiovascular disease or cancer. The summary relative risk estimates for the two outcomes, respectively, were 0.57 and 0.83. What was informative about this meta-analysis, however, was that those studies that showed the greater reductions in cardiovascular risk also showed the greater reductions in total cancer risk.

Postuma et al. suggested that this correlation may reflect a "healthy woman effect", the women at lowest risk for both cardiovascular disease and cancer being the ones who tended to take oestrogens. The correlation could also be explained by other shared cofounders, such as the style. Since it is clear that oestrogen itself cannot itself reduce the total risk of cancer, the spurious reduction in cancer risk that correlated with the reduction in the cardiovascular risk can only be explained by shared confounders, mostly in the same direction, across the studies. Incidentally, the findings of Postuma at al. also raise questions about the magnitude of the cardiovascular risk reduction, but that matter is beyond the score of this presentation.)

What is likely to be the future of meta-analysis? It appears that it is unlikely to go away, and for that reason some epidemiologists have argued that, rather than oppose it, a better approach might be to try to contain its excesses (Petiti 1994). I disagree. I think there is something profoundly amiss in the uncritical way in which epidemiologists, and indeed medical professionals in general, have allowed themselves to be seduced by the numerological abracadabra of meta-analysis. Perhaps the technique will succumb to its own absurdity, but if not, the next step in this surrealistic evolution will be the meta-analysis of meta-analysis, in which the meta-analysts will be totally divorced from reality, and totally surrounded by numbers without context. If anyone in this audience believes that development is far off, he should familiarise himself with the latest fashion of so-called "evidence-based medicine" and "systematic review" now playing on the internat (Feinstein 1995).

I would like to conclude by quoting A. Feinstein (1995). Feinstein and I have had our differences from time to time, but for once we are in total agreement: "the meta-analysis of non-randomised observational studies resembles the attempt of a quadriplegic person to climb Mount Everest unaided. "

References

1. Susser M (1991) What is a cause and how do we know one? A grammar for pragmatic epidemiology. Am J Epidemiol 133: 635-48
2. Boston Collaborative Drug Surveillance Program (1974) Reserpine and breast cancer. Lancet 2: 669-71.
3. Armstrong B, Stevens N, Doll R (1974) Retrospective study of the association between use of rauwolfia derivatives and breast cancer in English women. Lancet 2: 672-5
4. Heinonen OP, Shapiro S, Tuominen L et al. (1974) Reserpine use in relation to breast cancer. Lancet 2: 675-677
5. World Health Organisation (1980) Some pharmaceutical drugs. (IARC monographs on the evaluation of the carcinogenic risk of chemicals to man, vol. 24.) Lyon, France. International Agency for Research on Cancer 211-41. (Distributed by WHO Publications Centre USA, Albany, NY.)
6. Shapiro S, Parsells JL, Rosenberg L, et al. (1984) Risk of breast cancer in relation to the use of rauwolfia alkaloids. Eur J Clin Pharmacol 26: 143-146
7. Chalmers TC, Matta RJ, Smith H Jr, Kunzler AM (1977) Evidence favoring the use of anticoagulants in the hospital phase of acute myocardial infarction. N Engl J Med 297: 1091-1096
8. Longnecker MP, Berlin JA, Orza MJ, et al. (1988) A meta-analysis of alcohol consumption in relation to risk of breast cancer. JAMA 260: 652-656
9. Longnecker MP (1994) Alcoholic beverage consumption in relation to risk of breast cancer - meta-analysis and review. Cancer Causes and Control 5: 73-82
10. Shapiro S. (1994) Meta-analysis/shmeta-analysis. Am J Epidemiol 140: 771-778
11. Longnecker MP (1995) Re: "Point/counterpoint: meta-analysis of observational studies." Am J Epidemiol 142: 799-780
12. Shapiro S (1995) Dr. Shapiro replies. Am J Epidemiol 142: 780-781
13. Morris RD, Audet AM, Angelillo IF, Chalmers TC, Mosteller F (1992) Chlorination, chlorination by-products and cancer: a meta-analysis. Am J Path Health 82: 955-963

14. Alavanja M, Goddstein I, Susser M (1978) A case-control study of gastrointestinal and urinary tract cancer mortality and drinking water chlorination. In: Tolley RL, Gorcher H, Hamilton DH Jr, eds. Water Chlorination: Environmental Impact and Health Effects. 2nd ed. Ann Arbor, MI: Ann Arbor Science Publishers :395-409

15. Brenniman GL, Vasilomanolakis-Lagos J, Amrel J, Tsukasa M, Wolff AH (1980) Case-control of cancer deaths in Illinois communities served by chlorinated or non-chlorinated water. In: Tolley RL, Brungs WA, Cumming RL, et al. (eds) Water Chlorination: Environmental Impact and Health Effects. 3rd ed. Ann Arbor, MI: Ann Arbor Scientific Publishers 1043-1057

16. Howe GR, Hirohata T, Hislop G, et al. (1990) Review. Dietary factors and risk of breast cancer: combined analysis of 12 case-control studies. J Natl Cancer Inst 82: 561-569

17. Shapiro S (in press) Do trans fatty acids increase the risk of coronary heart disease? A critique of the epidemiological evidence. Am J Clin Matr

18. Hunter DJ, Spiegelman D, Adami H-O, et al. (1996) Cohort studies of fat intake and the risk of breast cancer- a pooled analysis. N Engl J Med 334: 356-361

19. Whittemore AS, Harris R, Itnyre J, et al. (1992) Characteristics relating to ovarian cancer risk: collaborative analysis of 12 US case-control studies. II. Invasive epithelial ovarian cancers in white women. Am J Epidemiol 136: 1184-1203

20. Harris R, Whittemore AS, Itnyre J, et al. (1992) Characteristics relating to ovarian cancer risk: collaborative analysis of 12 US case-control studies. III. Epithelial tumors of low malignant potential in white women. Am J Epidemiol 136: 1204-1211

21. Annegers JF, Strom H, Deckder DG, et al. (1979) Ovarian cancer: incidence and case-control study. Cancer 43: 723-729

22. Demopoulos RI, Seltzer V, Dubin N, et al. (1977) The association of parity and marital status with the development of ovarian carcinoma: clinical implications. Obstet Gynecol 54: 150-155

23. Newhouse ML, Pearson RM, Fullerton JM, et al. (1977) A case-control study of carcinoma of the ovary. Br J Prev Soc Med 31: 148-153

24. Wynder EL, Dodo H, Barber HRK (1969) Epidemiology of cancer of the ovary. Cancer 23: 352-370

25. Negri E, Franchesi S, Tzonon A, et al. (1991) Pooled analysis of 3 European case-control studies: I. Reproductive factors and risk of epithelial ovarian cancer. Int J Cancer 49: 50-56

26. Hartge P, Schiffman MH, Hoover R, et al. (1989) A case-control study of epithelial ovarian cancer. Am J Obstet Gynecol 161: 10-16

27. Cramer DW, Hutchison GB, Welch GR, et al.(1983) Determinants of ovarian cancer risk. I. Reproductive experiences and family history. J Natl Cancer Inst 71: 711-716

28. Nasca PC, Greenwald P, Chorost S, et al. (1984) An epidemiologic case-control study of ovarian cancer and reproductive factors. Am J Epidemiol 119: 705-713

29. Shapiro S (1995) Re: "The authors reply" to re: "Characteristics relating to ovarian cancer risk: collaborative analysis of 12 US case-control studies. II. Invasive epithelial ovarian cancers in white women." [Letter to the Editor] Am J Epidemiol 140:143

30. Shapiro S (1995) Risk of ovarian cancer after treatment for infertility. [Letter to the Editor] N Engl J Med 332: 1301

31. Whittemore AS (1994) The risk of ovarian cancer after treatment for infertility. N Engl J Med 331: 805-806

32. Postuma WFM, Westendorp RGJ, Vandenbroucke JP (1994) Cardioprotective effect of hormone replacement therapy in postmenopausal women: is the evidence biased? Br Med J 308: 1268-1269
33. Petiti DB (1994) Of babies and bathwater. Am J Epidemiol 140: 779-782
34. Systematic Reviews (1995) Chalmers I, Alltman DG (eds.) London: BMJ Publications
35. Feinstein AR (1995) Meta-analysis: statistical alchemy for the 21st century. J Clin Epidemiol 48: 71-9

Commentary on meta-analysis

Steven Goodman, Baltimore / USA

I do not know whether this will annoy S. Shapiro but I believe that contributions such as his will actually play a key role in ensuring that meta-analysis survives in epidemiology. I think his critique represents a necessary corrective action that we see in the development of any new method. Perhaps I should spare you my first thousand words and summarise his talk with a picture. I believe what he was essentially saying was that meta-analysis of observational studies gives you about as much guidance as the map does to the poor soul in this picture - a little man on his knees in the desert looking up forlornly at a map with no landmarks, with a dot in the centre labelled, "You are here". I believe this is what he was saying; that meta-analysis seems to tell us precisely where we are, but we have absolutely no idea whether we are right or wrong, or where to go from there.

My reaction to S. Shapiro's presentation was both strongly positive and strongly negative. I guess you could say that I have mixed feelings about it. First, I will talk about the things which we agree on and which I think represent the more important issues. The first issue that we agree on is that the evidence in meta-analysis is typically overstated. For two reasons, I would say that is true in both meta-analysis of randomised control trials and of observational data. The first reason is what I consider the poor handling of both qualitative and quantitative heterogeneity. I think he provided some very good examples of the failure to deal appropriately with qualitative heterogeneity. Meta-analysts who do not, or do not seem to understand the biology underlying their meta-analysis have a lot to learn from S. Shapiro's criticism. The second reason is based on a purely technical consideration which is my little soap-box. The use of p-values as evidential measures instead of likelihood measures creates an illusion that the evidence is much stronger than it actually is. We should be looking in meta-analysis for z scores of at least 3 and not 2. We should not be using conventional measures of significance if we are going to make supposedly definitive statements. By making just this one change, many spurious results that we see would be taken care of, but not all of them.

The second issue which I agree on with S. Shapiro is that in meta-analysis, all too often method takes precedence over biologic reasoning and creative hypothesis generation. This is true mostly in epidemiology. I do not think meta-analysis is unique in this respect. I offer two quotes along these lines. The first is from R.A. Fisher in 1958 when he said "I am quite sure it is only personal contact with the natural sciences that is capable of keeping straight the thought of mathematically-minded people. It is even worse I believe in this country (the USA) than in most.

Certainly there is grave confusion of thought. We are quite in danger of sending highly trained and intelligent young men out into the world with tables of erroneous numbers under their arms and with a dense fog in the place where their brains ought to be. In this country, of course, they will be working on guided missiles and advising the medical profession on control of diseases and there is no limit to the extent to which they could impede every sort of national effort." I think his point is self-evident.

The other quote is from AWF Edwards, a disciple of Fisher. He said: "What used to be called judgement is now called prejudice and what used to be called prejudice is now called the null hypothesis. It is dangerous nonsense dressed up as a scientific method, and will cause much trouble before it is widely appreciated as such." Both these quotes point to a generic problem in quantitative investigations of biological phenomena. Too many people tend to focus exclusively on the quantitative and lose sight on the biological. S. Shapiro and I both heartily agree on this point.

I believe S. Shapiro's point was that we need to look at studies individually instead of rushing to "add them up". Even if we agree that the synthesis of data gives us information that is less than the sum of the parts, I believe we can still say it is larger than the greatest individual part. The question is, how do we use the method of meta-analysis to appropriately assess the weight of this cumulative evidence? I do not wish to leave you without an equation since I know how you are hungering for some. Equation I is:

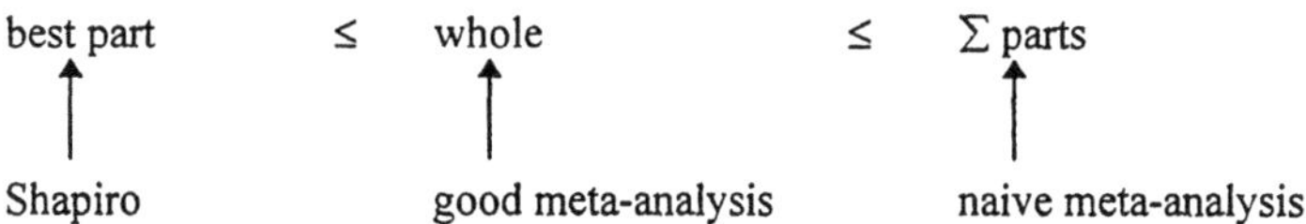

S. Shapiro obscures the question just posed by me in several ways. First of all, he frames the issue (in the paper submitted) as to whether meta-analysis is valid or invalid, which is false dichotomy. This polarises things and makes meaningful discussion difficult. It also avoids considering alternatives. Obviously, as with all experimental methods, either it is this method or nothing. The real questions might be answered in shades of grey, such as how useful is it?; how much information does it give us?, and is what it gives us better than what we will lose by using it? If the disadvantages of democracy were to be listed, these would be that it is messy, it is inefficient, it does not make the trains run on time. You would probably reject it as a stupid system. But when you compare it with the alternatives, most people would accept it as the best of many imperfect systems. We should evaluate meta-analysis the same way.

Dr. Shapiro's second diversion from the central question was by characterising meta-analysis as a mere calculation of weighted averages. That is not what meta-analysis is about. It is a process of *structured review* where taking the weighted average is only a small part of the whole process. We should not forget this. S. Shapiro correctly pointed out that the step of adding up studies is one that should be very carefully taken, but there was nothing in what he said that says we would be better off if we had not gone through a rigorous process before we reach that point of decision. He argues that the process before the summation should be more complex and rigorous than is often the case. I fully agree with his point of view.

Using the outline from the meta-analysis courses taught by me at Hopkins, I would like to illustrate that meta-analysis is not a mathematical formula. The steps for meta-analysis are: formulation of the review team, development of the question and protocol, collection of the literature, refinement of the question, qualitative and quantitative data extraction, qualitative synthesis, quantitative synthesis, sensitivity analysis, conclusions and dissemination. Quantitative synthesis is only a part of a very complex process. S. Greenland commented on this issue as follows: "The potential for abuse of a summary method should not impugn its proper use any more than, say, the abuse of morphine should lead us to withholding it from a patient in agony; the latter state perhaps induced by trying to make sense out of certain purely narrative reviews."

I will talk about narrative reviews later on. First I want to show something from the world of clinical trials. This example is from an article by E. Antman in JAMA in which the cumulative meta-analysis results of thrombolytic therapy after myocardial infarction were compared with what was found in textbooks and review recommendations at the time. Arguably, you have fairly definitive results first appearing in the early seventies but let us not quibble about the exact decade. There is little doubt that certainly by the early eighties definitive results could be found. In the mid-eighties we have the Isis trial going from 6,000 to 21,000 patients with no change in the point estimate. However, throughout the whole period virtually no mention of the therapy was found in books and reviews, except perhaps as an experimental one. You cannot imagine a stronger motivation for the literature to cite this therapy when it involves an issue of life-and-death and the evidence is compelling. Yet, for reasons unknown, it was not reviewed in the summary literature. It seems to be that a serious problem exists with what is included in systematic reviews and how it is synthesised, which makes me a little distrustful of the process implicitly proposed by S. Shapiro. That is, that we should depend on basically the same sort of thing that we have had in the past, a sort of review by a designated "expert" often tied together with a very good biologic hypothesis, but not necessarily with a complete knowledge on what is in the literature.

Another issue I wish to illustrate is the impact that good meta-analysis can have. Time is too short to tell the complete story here. The issue concerns the question of whether women undergoing premature labour should be given steroids to accelerate the maturation of the fetal lung; to reduce respiratory distress syndrome (RDS), and possibly even reduce mortality among premature infants. To shorten the story to just two sentences, this is a treatment that, as early as 1980, proved to be effective with no serious adverse effects. A meta-analysis was done in 1980 and a more complete one in 1990. Yet, in 1994 a NIH consensus conference on the subject had to be convened because only approximately 10% - 20% of practitioners in the United States of America were actually using this treatment. Although all the evidence can be found in the literature in the form of randomised control trials, for some reason this weight of evidence was not appreciated by the medical community.

These are the results of a NIH consensus conference which took place in 1994 on this issue, which was spurred largely by the 1990 publication on a meta-analysis. This is actually the basis for the logo of the Cochrane Collaboration: "A recent meta-analysis concluded that steroid administration prior to pre-term delivery is associated with a large reduction in the incidence of early death, respiratory distress syndrome, intraventricular haemorrhage and necrotising enterocolitis". Only 12% - 18% of eligible women who delivered in the USA were getting this treatment in 1994, arguably at least 15 years after it appeared in the literature. The report concluded that the use of antenatal corticosteriods for prematurity was a "rare example of a technology that yields substantial cost savings in addition to improving health". This reinforces the point that we do not always get a good synthesis and dissemination of what is out there, even with the strongest stimuli, and that a meta-analysis is sometimes necessary to attract people's attention.

I would like to present some examples of meta-analyses of observational data. I have two pairs of articles here. One pair is on the effects of oral contraceptive use on breast cancer and the other is on estrogen replacement and breast cancer. They will not be reviewed in detail here. You can analyse them systematically because of their stated protocols, which include the databases searched, the dates of the studies, the non-database sources for the studies, the language of the studies that they included, their exclusion criteria, and what studies they found. Each had a variety of design components to be looked at: whether the data was grouped or individual; dates of the study; sources of control; sources of cases; data collection method, and matching and adjustment factors. The analytic methods are based on issues like duration, dose, age at birth, pregnancy, family history, age, latency, parity, drug type, benign disease, and other design features.

To my delight, and S. Shapiro's as well, each pair of meta-analysis came to different conclusions using almost the same studies. One of each pair essentially showed that heterogeneity between studies prevented clear conclusions, whereas the other study drew fairly specific quantitative summary statements. For example, one said "... differences in study design have substantial effects that obscured clear conclusions, ..." whereas design features were not included in the other meta-analysis. I would contend that it is the very process and protocol of meta-analysis that allows you to examine these and figure out how and why they came to these different conclusions. As a reader, you have access to the same information as they, so you can judge for yourself. In this situation, I was on the side of those who said that heterogeneity was too great to justify pooling.

Because I do not think that we can consider meta-analysis alone, a very crude and simple comparison can be made between a narrative review and a meta-analysis.

1. *Completeness?:* Meta-analysts certainly try to be, but in the narrative reviews one has no idea whether everything relevant has been included in the literature.
2. *Clear criteria for study identification and study eligibility?:* This can be found in meta-analysis but not in narrative review.
3. *Comprehensive description of the studies?:* Yes, in a meta-analysis. No, in a narrative review, although supposedly there could be.
4. *Can readers independently criticise the data, methods and conclusions?:* I will use the medical way of scoring. On a scale of 1 to 4 (highest), I give meta-analysis a 4 for this kind of explicitness, which is expected of all good research, and a 2 for a narrative review.
5. *Is there a coherent synthesis based on biological knowledge?*: I think many narrative reviews do much better on this point, so they score a 4 here and meta-analysis a 2. However, there is absolutely no reason why meta-analysis cannot do exactly the same thing and better. Meta-analysis have discussion sections where studies can be included, excluded or down-weighted using exactly the kinds of criteria suggested by S. Shapiro. Quality scoring done on purely operational factors is a great problem. There has to be a second level of evaluation having to do with coherence with some biological explanation, like in M. McClure's paper on "deductive meta-analysis". There have been a number of meta-analysis where analysis of subsets based on biologic thinking, such as the time period when anticoagulants after myocardial infarctions were given, resulted in different findings.
6. *Inappropriate weight of the results?:* Narrative review - 1, meta-analysis - 4. I completely agree that in a meta-analysis there is a tendency to give the results inappropriately high weight. This problem needs to be solved. The narrative review is less subjected to this. Narrative reviews have a problem in that for some reason or other, they are unable to get the medical community to appreciate the full power of the synthesised evidence.
7. *Biased results?:* We do not really have a gold standard so it is very difficult to judge. This is why something resembling a "scientific method" is used to study nature. Meta-analysis come closest to this. Informal methods are less open and amenable to corrective criticism such as that which S. Shapiro has directed at certain meta-analysis.

One quick point is that we have to consider the dangers of using meta-analysis. What are the consequences and what are the problems? If the consequence is action or policy or treatment without further study, perhaps then there is a problem. However, from some evidence that will not be presented here, meta-analysis do stimulate either larger or better studies of the same or different design, or sometimes even another meta-analysis.

The real questions to be confronted are:

1. *How do we decide whether studies are combinable?* This must be recognised as an issue even when the odds ratios are equal. It is not necessarily true that when two studies have odds ratios of 1.5, this presents a *prima facie* evidence that one can "add the studies up" and claim stronger aggregate evidence. I definitely agree with S. Shapiro on this and would say that the method which should be used to judge this should be the S. Shapiro method, which is using more biological coherence.
2. *If we combine the studies, how can we and how do we reflect the extent to which the qualitative heterogeneity should affect the inference?*
3. *How do we educate the profession and the media about the proper response to meta-analysis?*

We have the advantage of seeing how S. Shapiro would respond to some of the points made by me. He says that a narrative review allows for more creativity and biological intuition. I do not understand why this is better than a good meta-analysis. The point has been made by others here: if you reject meta-analysis, do you reject all observational epidemiology? He stated in writing that this misses the point; that individual studies should lead to better individual studies with fresh approaches. My response to this is that I do not see why that cannot be true for meta-analysis as well. At least it is assessable and open to criticism.

S. Shapiro said that we should discredit bad science (i.e. meta-analysis) because it is bad science and cannot be accepted because it is the current fad. My question is, what is the *good* science that he proposes? Should not the review and aggregation of the research be done scientifically? By this, it means that it be done openly, with clear criteria and letting all audiences have access to the data and the process. Where people have said that it is useful, S. Shapiro said: "where are the studies that show its strength?". I suggest that they are the very ones that he has shown us today. He has shown how the explicitness of the meta-analysis has allowed him to target his criticisms and theoretically show the way to an improved synthesis.

Finally, S. Shapiro stated that too many epidemiologists are willing to equate data aggregation with the truth. For that, I have an unassailable rebuttal. I agree.

Discussion remarks on meta-analysis

Colin Begg, New York / USA

Basically, I agree with most of what S. Goodman said, that meta-analysis can be a very useful tool. I also agree with much of what S. Shapiro said, except the conclusions that meta-analysis cannot be useful. basically. To briefly state my point of view: I am a great believer in the scientific method, as I am sure all of you are too. To me, the essence of the scientific method is its truth-seeing objective, and that encompasses the ability to reproduce experiments. In the context of the literature, it means that all of us who want to interpret the literature should have access to the facts. And part of what I am saying relates to what was discussed in this conference about publication bias. I feel that in any publication of a study, you want to separate the facts about the study - how it was designed and the data that were collected, from the opinions or the spin that you want to put on the results. I think that one of the positive features that use of meta-analysis is promoting and will continue to promote is the factual reporting of data, as opposed to selective reporting of results that appear more positive than they would if you had access to the complete data. What is the role of meta-analysis? A meta-analysis is really not an individual study, and I think we can turn to what has happened in the randomised trial area for some guidance in this. S. Shapiro mentioned that one of the great successes of meta-analysis was the review of the adjutant trials on breast cancer. Essentially, what happened there was that there was a large literature of trials that were seemingly contradictory, but when they were all brought together in the review, a consensus was formed. I think it is a broadly accepted consensus that adjutant therapy works. So really the proper role of meta-analysis is consensus-building and its success or failure in that role will be evident after the meta-analysis is published and disseminated. You will either succeed in building a consensus or you will not. It would be great if you succeed in building a consensus, as in the treatment of breast cancer. If you do not succeed in building a consensus, then the evidence is simply not persuasive enough at this point and more studies have to be done. I also think that, although it will undoubtedly be harder to build a consensus in the area of observational epidemiology, the proof is in the pudding. You have to try, and if you do not succeed in building a consensus, then the issue need not be resolved. Many of the examples that were presented by S. Shapiro show that the style of analysis is really inadequate as it is currently practised. I agree with S. Goodman that we need to improve the style of analysis. We need to focus more on evaluating in a quantitative way the possibilities of bias, to study heterogeneity and so forth, and not go right in to a simple summarisation of the data, where apples and oranges are combined. From reading the work of people really working on meta-analytic methodology, I feel that it is well accepted, that methodology needs to be improved. One should not just obtain a summary estimate in an unthinking way.

A final point, with regard to the primacy of individual studies versus aggregation as in a meta-analysis, is that the popularity of meta-analysis is really a historical inevitability at this time. Fifty years ago, scientists could usually mount a study and assume with some confidence that it was probably a unique study. So, the evaluation of the results at the end were very important and unique, and I think our style of data analysis reflects that. In fact, the scientific milieu has changed enormously, such that there are few if any unique studies nowadays, and lots of different studies have been published on any given topic. The whole focus of attention has changed from evaluating the aggregation of information from multiple studies. Now this is not a simple thing to do, as the examples presented show, but this is where it is at. We have to improve our ability to accomplish this effectively. S. Shapiro said to me earlier that he thought meta-analysis would be discredited and people would not do them any more, at least of observational studies, but I just do not see that happening. I think that the review of the literature on a specific topic and the focus on the quantitative nature of that review are here to stay and we really have to work to improve the methods. I would like make one last point and that is that I cannot resist responding to my name being at the bottom of the league table in that alcohol example. In fact, the article we published was a letter to the editor. When people do these quality weights, which, I agree with S. Shapiro, are largely worthless other than as a qualitative evaluation of the studies, really to a large extent they represent the completeness of the reporting of the methodology of the study. However, when you write a letter to the editor, you do not really spend a lot of time listing all the details of your methods, so it is not unreasonable that you would end up at the bottom of the league table. This was not mentioned in the article.

Publication bias

Kay Dickersin, Baltimore / USA

I would like to begin by providing a context for my discussion on publication bias: meta-analysis. As an epidemiologist, I have been particularly interested in data collection issues in relation to systematic reviews and meta-analysis. A meta-analysis is an observational study, where the study population comprises of studies and not people. As such, it has many of the same problems that any observational study has, even when the studies included are randomised trials. For example, the results of a meta-analysis may not be valid if the selection of studies to be included is biased. Publication bias is a special case of selection bias, observed when positive results are differentially published compared to negative results. A meta-analysis, or any systematic review, depends on data from published reports.

A goal in performing a systematic review or meta-analysis is either to achieve complete ascertainment of all studies done relating to the chosen topic, or to identify an unbiased sample of all studies. Since it is not clear how to take an unbiased sample when the denominator is unknown, it is generally recommended that investigators attempt to identify all studies in a given area. Unfortunately, one never knows if one has found "everything". Consequently, a fair amount of methodological work has been done to identify the potentially problematic areas.

First, identification of all relevant published reports is not easy. The precision (equivalent to positive predictive value) and sensitivity of electronic searching (for example MEDLINE) have been estimated and found to be quite low (Dickersin et al. 1994). Precision can be improved, but with serious consequences for sensitivity. Thus, the average investigator is faced with the choice of, either increasing the number of relevant studies retrieved, at the expense of having to review literally thousands of citations, or retrieving a smaller proportion of relevant studies, but only needing to review a manageable number of citations. The result of this dilemma is that incomplete ascertainment is almost certainly a factor in any meta-analysis. What is not clear is whether it has ever biased the findings.

Second, as noted earlier, selection bias is a particularly worry in a meta-analysis. Not all study findings are published and there is strong evidence of an association between positive results and publication (Dickersin et al. 1992, Dikkersin & Min 1993, Easterbrook et al. 1991, Dickersin & Min 1994). Unfortunately, there are no satisfactory methods available to identify unpublished studies (Hetherington et al. 1989). Other types of selection bias may also be operating. For example, some investigators choose to limit their meta-analysis to reports published in English or years of publication. These restrictions may also risk the validity of the review.

Different types of bias are possible, for example, the tendency to publish results that conform to cultural attitudes or traditions. A. Vickers and his colleagues have compiled a database of clinical trials of acupuncture and compared the reported findings by country of origin. They found that almost 100% of trials from Eastern Asia and Eastern Europe were positive studies, while approximately 50% of reported trials from North America and elsewhere in the Western world were positive. In this case, either there is a cultural publication bias favouring a particular result, or practitioners in the West are not as proficient at acupuncture as those in the Eastern countries (A. Vickers' personal communication to K. Dickersin, 4 September 1996).

Given the context of meta-analysis, I would like to explore publication bias a little further. First, what do we know about publishing practices in general? There are 11 studies that followed abstracts reporting the results of various types of studies - including trials and observational studies - from various fields, to determine the proportion published (Scherer et al. 1994). On the whole, only about 51% (95% CI = 45-57%) of studies that are presented in abstract form ultimately end up being published. This indicates that the published literature is not representative of all research undertaken. Figure 1 summarises the full publication rates over time for 8 studies for which data were available.

Figure 1
Percentages of total abstracts published over time, calculated for individual studies.

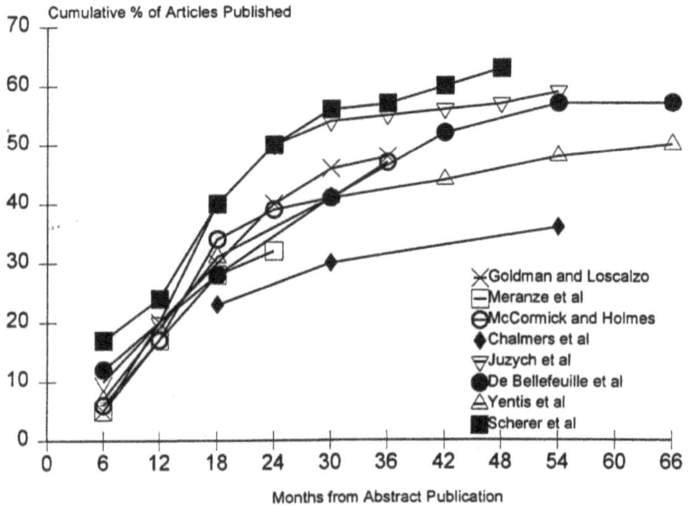

Scherer R, Dickersin K, Langenberg P. Full publication of results initially presented in abstracts: A meta-analysis. JAMA 272:158-62, 1994. Correction JAMA 272:1410, 1994.

Four studies that have prospectively examined publication and publication bias have been published to date (Dickersin et al. 1992, Dickersin & Min 1993, Easterbrook et al. 1991). Our group followed two populations at Johns Hopkins. All studies were approved in 1980 by the institutional review board at the Medical School ("MED") and all studies approved that year by the institutional review board of the School of Hygiene and Public Health ("PH"). We also followed all clinical trials funded by the National Institutes of Health in 1979 (Dickersin & Min 1979), excluding those funded by the NCI, for reasons to do with the different ways in which cancer studies are designed and conducted (e.g. treatment comparisons may change mid-course). Studies in these three cohorts were followed in 1988 to learn about the study findings and publication status. The fourth study followed in 1990. All the projects were approved by the Central Oxford Research Ethics Committee (COREC) in 1984-87 (Easterbrook et al. 1991). This study used data collection forms modelled after forms used in the other studies, allowing for straightforward combination of results. The John Hopkins and Oxford cohorts included a mixture of clinical trials and observational studies, although the vast majority of studies initiated at the School of Public Health were observational. As noted earlier, the NIH cohort comprised of studies the NIH defined as clinical trials, including nonrandomised studies.

These studies showed that between 52% and 93% of initiated studies were published. Projects initiated in Oxford had the lowest publication rate and trials funded by the NIH had the highest.

Table 1
Publication rates of initiated studies

Source	% published
Johns Hopkins "MED" (1980 studies)	81%
Johns Hopkins "PH" (1980 studies)	66%
NIH (1979 clinical trials)	93%
Oxford COREC (1984-87 studies)	52%

For studies initiated at Johns Hopkins, there are only minor differences in risk factors for publication between studies conducted at the Medical School and studies conducted at the School of Public Health (Dickersin 1992). In both populations, significant results were associated with publication, but this association was not statistically significant for PH. Sample size did not appear to be associated with publication, nor did multicentre status, or having a comparison group. Trials were no different from observational studies in terms of reaching publication. External funding was the only factor other than significant results that was significantly associated with publication. Similar results have been obtained in every study that has examined this factor. Neither sex nor rank of the PI was associated with publication.

124

Risk factors for publication were similar for NIH clinical trials funded in 1979 (Dickersin & Min 1993). Again, significant results were significantly associated with publication (OR=7.04; 95%CI = 1.90 to 26.16), despite the fact that so few studies remain unpublished.

Given the similarity in design and data collection, we felt that combining the results of the two Hopkins studies, the NIH, and the Oxford studies in a meta-analysis would be useful for obtaining an estimate of the true size of the association between significant results and publication. We estimated a combined odds ratio of 2.88 (95% CI = 2.13 to 3.89)(Dickersin et al. 1993).

Figure 2
Meta-analysis of four studies examining the association between significant results and publication: unadjusted odds rations and 95% confidence limits

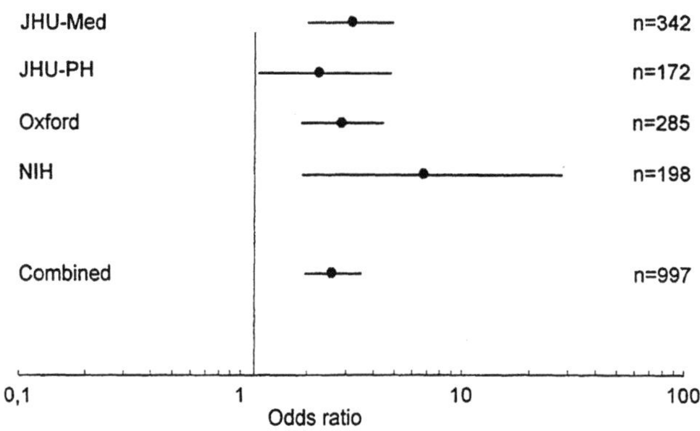

Dickersin K, Min Yi. NIH clinical trials and publication bias. Online Journal of Current Clinical Trials 1993 April 28 (Doc No 50)

When we asked investigators of completed unpublished JHU studies (PH and Med combined) why they had not published, the vast majority (95%) replied that they had not submitted a manuscript: their reasons included that they felt their results were not interesting (31%); publication was not an aim (14%), and the study has design or operational flaws (34%) and other reasons (21%). Only rarely had a manuscript been submitted and rejected. Our study of NIH trials and the Oxford study had similar findings: the main reason study results remain unpublished is that investigators fail to write up their findings, and not that editors reject "negative" or other results.

Thus, we have established strong evidence for publication bias, whether one is interested in observational studies or clinical trials. This could have a major impact on a meta-analysis. Stewart and Parmar (1993), for example, compared meta-analysis using individual patient data compared with meta-analysis using data available in the literature and found that not only did several studies never get reported, but those that did failed to report all the variables investigated. Variables not mentioned in published reports tended more often to be those for which there was no significant association with the outcome in question.

An example of a current area of interest with regard to a potential publication bias is the possible association between abortion and breast cancer. It is easy to imagine that positive findings related to this controversial topic might be differentially published. A collaborative group of investigators representing over 50 observational studies of breast cancer is currently pooling individual patient data to examine risk factors for breast cancer (Collaborative Group on Hormonal Factors in Breast Cancer 1996). Although they have not published a report examining the association between breast cancer and abortion, this group is in a key position to learn whether the published findings reflect what is actually known.

In summary we know that perhaps as few as one half of all studies initiated are published in full. We also know that studies with positive results are more likely to be published than studies with negative findings. The failure to publish usually rests with the investigators, not the editors. These findings are of concern to us generally because of the potential bias that is introduced into any formal or informal data synthesis when we rely on published results.

There is another implication of these findings that is particularly important and disturbing. Patients who participated in these studies probably signed a consent form stating that by their participation they will be contributing to medical research. Therefore, any failure to publish study findings is failing to honour the trust placed with the investigators by the patients who agreed to participate in a study. Failure to publish could even be considered unethical.

What mechanisms can be put in place to anticipate and counteract the problem of publication bias? Many of us have advocated for a number of years that we need a register or registers of initiated studies, to provide us with a denominator of all studies undertaken (Dickersin 1992, Ad Hoc Working Party of the International Collaborative Group on Clinical Trial Registeries 1993). There are many registers of clinical trials, though none is comprehensive. The most successful model in the US is for AIDS (Dickersin 1992). Cancer trials registers are also in existence in the US and Europe. The Cochrane Collaboration (Bero & Rennie 1995), a world-wide effort launched in 1992, aiming to prepare and disseminate systematic reviews of intervention related to health care, is in the process of developing a register of published reports and has as its long term goal registration of all initiated trials.

Something also need to be done to register observational studies. I would propose that any register of initiated observational studies must also include a list of the variables for which data were collected. This will be especially important for examination of environmental and occupational exposures, and the presence of various genetic markers, and provide the means for future collaboration and data-sharing. Epidemiologists should consider a "Cochrane"-like project to systematically collect and review associations among many risk factors and outcomes examined over time. Although this would be a monumental undertaking, failing to take this challenge puts us at risk of losing control of all that we have learned so far.

References

1. Dickersin K, Scherer R, Lefebvre C (1994) Identification of relevant studies for systematic reviews. BMJ 309: 1286-1291
2. Dickersin K, Min YI, Meinert CL (1992) Factors influencing publication of research results: Follow-up of applications submitted to two institutional review boards. JAMA 267: 374-378
3. Dickersin K, Min YI NIH clinical trials and publication bias. Outline Journal Current Clinical Trials April 28, 1993.
4. Easterbrook PJ, Berlin JA, Gopalan R, Matthews DR (....) Publication bias in clinical research. Lancet 337: 867-872
5. Dickersin K, Min N (1994) Publication bias: The problem that won't go away. In: KS Warren, F. Mosteller (eds), Doing more good than harm, the evaluation of health care interventions. Ann NY Acad Sci 703: 135-146
6. Hetherington J, Dickersin K, Chalmers I, Meinert C (1989) Retrospective and prospective identification of unpublished controlled trials: Lesson from a survey of obstetricians and paediatricians. Paediatrics 84: 374-380
7. Scherer R, Dickersin K, Langenberg P (1994) Full publication of results initially presented in abstracts: A meta-analysis. JAMA 272: 158-162. Correction JAMA 272, 1410 (1994)
8. Stewart LA, Parmar MK (1993) Meta-analysis of the literature or of individual patient data: is there a difference? Lancet 341, 418-422
9. Collaborative Group on Hormonal Factors in Breast Cancer (1996) Breast cancer and hormonal contraceptives: collaborative reanalysis of individual data on 53,297 women with breast cancer and 100,239 women without breast cancer from 54 epidemiological studies. Lancet 347: 1713-1727
10. Dickersin K (1992) Why register clinical trials? - revisited. Controlled Clin
11. Bero L, Rennie D (1995) The Cochrane Collaboration. Preparing, maintaining and disseminating systematic reviews of the effects of health care. JAMA 274:1935-1938

Concluding remarks

Richard Doll, Oxford / UK

I have two firm personal conclusions. One is that everyone at this conference could do a summary better than I could, and the other is that if we need a summary at all, we need two - one from an epidemiologist and one from a biostatistician. Apart from those two firm conclusions, I have a few others to which I hold less strongly.

The first is that the objective of this conference has been, I think, not the assessment of small effects as described, but the assessment of small relative risks. In fact, it is not always difficult to assess small absolute effects. They can often be very easy to detect, if the effect is normally exceptionally rare. Think of the adverse reactions to drugs that have been detected with quite small numbers, when the reactions very rarely or never occur under any other conditions. Think of peculiar occupational risks like that of angiosarcoma of the liver in vinylchloride workers; a risk which in other circumstances affects one in ten million people per year. Nor was it difficult to recognise that vaginal adenocarcinoma in young women was liable to occur when the mother was given stilbestrol during her pregnancy. These were all important discoveries, as they prevented the widespread use throughout the community of hazardous materials, and they were not difficult to make. The important subject with which we are concerned are the small relative risk and the small absolute risk of a common disease extrapolated from larger risks at higher dosage.

An example of the latter is the public concern about the effects of living near a nuclear reactor. Epidemiologists have to tell the community what that is. We do tell them; but we do not succeed in persuading them what the risk is! We tell them, as a result of making observations of the effect of higher exposures, making an estimate of the nature of the dose-response relationship and then extrapolating, in conjunction with biologists who tell us whether the type of extrapolation we make is sensible. We can do more. We can check to see that our estimate of the small effect has not been too small and this has been done in relation to the effect of fallout from nuclear bomb testing. We have shown that the effect is certainly not materially underestimated by the extrapolation techniques that we use. Unfortunately, some other problems are much more difficult. Consider lead and mental defect. There is no way in which we can actually measure the effect on mental defect of the very small amounts of lead in the blood which, it has now been decided, should be the upper permissible public limit. We can estimate the effect only by making observations at higher exposure and extrapolating. That is fair enough in the case of cancer, where a mutation is involved, as we have biological reasons to think that mutations are produced in linear proportion to doses down to vanishingly small levels. In relation to mental defect it is much more uncertain, for the mechanism is unclear and the evidence for the nature of the dose-response relationship is not compelling. Unfortunately, it is a type of problem that we have to face up to as epidemiologists. We cannot avoid the responsibility of assessing the reality of the risk although the relative risk may be well below 1.5, if the exposure is really common. Other examples are the risk of asthma as a result of exposure to the exhaust fumes from cars, the risk of myocardial infarction from different types of coffee, and the risk of hormone replacement therapy which we must try to settle in the next few years. Then there is the possibility of another small benefit - aspirin, which may reduce the risk of large bowel cancer by some 10%. This is something that I believe we should try to get straight, because it is perfectly practical for very large numbers of people to take small doses of aspirin. Large number of people are already taking it in the United States because they think it might protect them from myocardial infarction. The habit might even spread to Europe, if we show that small doses reduce the risk of carcinoma of the colon as well.

In our discussion, we began by considering the evidence that would lead us to conclude that an association reflected cause and effect. We referred to the guidelines set out by Hill, Lilienfeld and others. I do not think we can improve on them. We have not discussed them subsequently and perhaps it is just as well because it is doubtful that anything more useful can be said about them than has already been mentioned several times in the past. Namely, that the determination of causality in the absence of controlled experiment is a matter of subjective judgement and that the guidelines are there just to help us reach it. Every case has to be considered on its own merits. There are no simple rules that enable us to say the association meets this, that and the other criterion, therefore, it is a causal association, QED. The guidelines are no more than was originally claimed for them, namely, they help us to think. None is conclusive; only one, a temporal relationship, is essential. A number of important decisions have been made on the basis of one study. B. Hill referred to his example on occupational hazard of cancer of the nose, to which his attention was drawn by the industry concerned. He showed that ten cases of nasal sinus cancer occurred in process workers in a nickel factory, when he would have expected from national data less than 0.1. There was no other biological evidence of a carcinogenic hazard from nickel or the nickel salts to which the workers were exposed. The evidence was, however, quite sufficient for the factory to take action to reduce the exposures to a minimum. In fact, the disease was eliminated locally. Unfortunately, it was not eliminated in some other places, as the evidence was not widely published. An example of a relative risk of 1.3 due to cause and effect, I decided, is the increased risk of myeloid leukaemia in cigarette smokers (Doll, 1996). It was not particularly difficult to reach this decision because we have evidence of leukaemogens in cigarette smoke (namely, radioactive polonium and benzene) and plenty of other evidence of other carcinogens circulating around the body. We did not find a strong association and had only weak evidence of dose-response relationship. However, we had evidence that, when people stopped smoking, the risk diminished. The evidence from several studies was consistent within the limits of random variation. We had the rather unconsidered guideline of analogy to guide us, what S. Shapiro would describe as "the company it keeps". We also observed that the association was specific for myeloid leukaemia, chronic lymphatic leukaemia being quite unrelated to smoking. Another example where I believe we have been able to reach a conclusion of causality is the use of oral contraceptives and the subsequent risk of breast cancer, although the risk is certainly down among the levels we are talking about. How we have been able to reach such a conclusion will be referred to shortly.

Before considering the guidelines that help us to reach a sensible decision, there are three explanations for the observations that we have to eliminate, which everyone knows about, namely, bias, confounding, and chance. Bias is most effectively eliminated, generally speaking, in cohort studies, although S. Shapiro did warn that there could be situations in which bias could affect cohort studies; but they occur only seldom. Statisticians tend to combine confounding with bias and call confounding a form of bias. Personally, I think it is helpful to distinguish them and to think of bias as something that is inherent in the conduct of the study, whereas confounding is something that occurs independently in society. Biases we have heard, is of two general sorts: selective bias in determining how we have chosen our subjects and controls, and information bias, which can enter in a whole variety of forms. Bias always has to be considered as a possible explanation of any association that we observe. It is difficult to avoid entirely, but as our discussions have shown, there are various ways in which it can be reduced.

Confounding becomes more important the smaller the relative risk and is seldom likely to account for relative risks greater than three. Nevertheless, it can do so occasionally. We have an example of a tenfold relative risk due to confounding: namely the mortality from cirrhosis of the liver in heavy cigarette smokers, which in one study (Doll et al. 1994) could be attributable to the concentration of heavy alcohol drinkers among the heavy cigarette smokers. We heard in the discussion that confounding could be defined as an association between two factors which would have allowed for experimentally, if one had had control over the situation. In other words, it occurs with factors that you have reason to believe are related to the risk of the condition under study. I was pleased to hear this definition because there are so many papers that end nowadays by saying "of course, we cannot conclude that this association is causal because it might be due to confounding with some unknown cause". That is a refuge of the destitute. We should have the courage to say : "We have looked at every possible cause of this condition that we know of and we cannot find any evidence of confounding", or alternatively: "We have seen some evidence of confounding, have allowed for it, and the relationship persists, and we conclude that it is causal". Consider cancer of the brain and its relationship to electromagnetic fields. When the association is reported in an occupational study, someone will certainly say: "Maybe it is due to confounding" - but what is it due to confounding with? I wish I knew of any factor that causes cancer of the brain, that exposure to electromagnetic fields could be confounded with! Unless someone can suggest a confounding factor to consider, the preferred interpretation of an association not attributable to bias or chance should be a causal one - if the guidelines make it reasonable.

We did discuss the problem of allowing for confounding when there was some factor which we knew affected the disease, but I do not think we really dealt with the problem of our incompetence in adequately eliminating it. I did suggest that if a rather crude allowance for confounding halved the Chi^2 value for the difference between the exposed and unexposed, then one should be a little hesitant to conclude that the other half would not also be eliminated by confounding if it could be allowed for better. Socio-economic status is a particularly difficult factor to allow for. Several studies recently have shown that *helicobacter pylori* infection is associated with myocardial infarction - a threefold risk of myocardial infarction being observed with positive evidence of *helicobacter pylori* infection. Such an infection is very closely associated with socio-economic class. An attempt to allow for it halves the apparent relative risk. Should *helicobacter pylori* infection still be considered a candidate cause of myocardial infarction, or have we just failed to allow for socio-economic conditions adequately, which in many countries are now an important contributory factor in causing the disease? Problems of this type deserve further consideration and examination by statisticians.

Then there is chance. We may all think we have known how to eliminate chance for years but of course we have not for many of the small relative risks that have been reported. Meta-analysis has been proposed as a means of getting rid of it altogether. I think we agreed that a routine meta-analysis of published results is quite useless and may be misleading. But I believe, and I think most participants would agree with me, that there is no escaping the need to assess quantitatively and qualitatively what the results of a whole lot of studies imply. In my view, the most certain way of getting a sensible answer is by what I call "collaborative re-analysis": that is, by bringing together for the purpose, all the epidemiologists who have studied a subject. One of the first things they then have to do is to enquire whether publication bias could have seriously affected the results and to seek any unpublished results that may exist: "Having overcome this hurdle, they all then need to agree on how to categorise their data so that they can be sensibly combined. Biases in different studies must be discussed and, if necessary, some studies must be rejected. I believe that some really useful results can then be obtained. I would like to be able to report the results of a collaborative re-analysis of the studies of oral contraceptive use and breast cancer that has been carried out in this way in Oxford. Sixty investigators were asked to collaborate and 51 did so by presenting all their data in an agreed form. Eight were unable to find their data because the studies were done so long ago that these were largely lost. One investigator declined to participate. Having been recorded centrally, the results were then printed out and sent back to the collaborators to check that they agreed with the way their findings were treated. The result produced a clear answer but one of the conditions for a collaborative re-analysis of this sort is that the results are not published in the name of any one individual, but in the name of all the collaborators, and that nobody should be informed about the results until the collaborators themselves have agreed on the final report. So it will take some time before the results of the collaboration can be seen. Inevitably, there will be overviews between these two extremes, between the collaborative re-analysis, which I hope everyone will agree is a worthwhile procedure, and the often useless routine analysis of published data without any attempt to prune them, find out whether any material data are unpublished, and assess the validity of the individual reports.

As to individual epidemiological studies, there are clearly some improvements in technique that can be used in the future. We have not mentioned the nested case-control study; possibly because everyone appreciated that it was probably the best technique for the future, not in all circumstances obviously, but in very many. It has not been used much in the past but will, I think, be used much more often than it has been before. Then we shall have to concentrate on obtaining more reliable measures of exposure, using objective measures whenever we can.

There is another technique that has not been used much in the past, but which I hope will be used more in cohort studies. This is the technique for allowing for dilution bias. Two examples of which this has been done, with valuable results, are in studies of the risk of myocardial infarction in relation to cholesterol level and the risk of cerebrovascular stroke in relation to blood pressure. In the former, a large cohort of people whose blood cholesterol levels were measured was divided into deciles according to the level and then re-examined six months later. Because of the tendency for regression to the mean, what was then found was that the mean value of the lowest decile was raised and the mean value of the highest decile lowered, with the consequence that the dose-response relationship was much steeper and a better estimate of the risk was obtained from the cohort study than would have been obtained if just a single observation on everybody had been used. It is not, of course, necessary to re-examine everyone in each decile as random samples will do, but the larger the sample, the better and, in practice, it may be administratively simpler to re-examine everyone. In some studies it pays to select groups at the extreme ends of a distribution for study. This was the way that the relationship with lung cancer was first demonstrated clearly, when E. Wynder and I took considerable trouble to define the lifelong non-smoker, which had not been done in the past. In cohort studies, it may sometimes be sensible to determine the distribution of exposures and then concentrate the study on populations that are at the extremes of the distribution. Unsatisfactory though it is, we have to accept that there is often no alternative to making observations on groups with relatively high exposures and then extrapolating to risks at low doses with biological advice about the likely nature of the dose-response relationship.

132

My final and most important conclusion cannot be controversial. That is, that epidemiologists need to have an expert biometric statistician working with them and expert biometric statisticians need to work with epidemiologists and not sit back by themselves and comment on results. That is obvious but the meeting has made this even more clear and brought out the tremendous importance of having expert biostatisticians and epidemiologists working closely together the whole time. We need to be constantly seeking to improve our methodology and we need, in my opinion and I believe also in the opinion of the group, to undertake collaborative analysis. If we can do that, I do not think we need to be put off by the idea that relative risks of the order of 1.5 are capable of being established. We can establish them and we can determine in appropriate cases that they indicate causality.

Reference

Doll R, Peto R, Wheatley K, Gray R, Sutherland I (1994) Mortality in relation to smoking: 40 years' observation on male British doctors. Brit Med J 309: 901-911

Acknowledgements:

The editors would like to thank B. Sonnenberg, P. Ross, U. Ellert and Y.M. Chew for their assistance in the preparation of the manuscript.